Dugald Clerk

The Gas and Oil Engine

Third Edition

Dugald Clerk

The Gas and Oil Engine
Third Edition

ISBN/EAN: 9783744742986

Printed in Europe, USA, Canada, Australia, Japan

Cover: Foto ©berggeist007 / pixelio.de

More available books at **www.hansebooks.com**

THE
GAS ENGINE

PRINTED BY
SPOTTISWOODE AND CO., NEW-STREET SQUARE
LONDON

THE

GAS ENGINE

BY

DUGALD CLERK

THIRD EDITION

NEW YORK
JOHN WILEY & SONS, 15 ASTOR PLACE
1890

PREFACE.

In this work the author has endeavoured to systematise the knowledge in existence upon the subject, and to explain the science and practice of the Gas Engine in a way which he hopes may be useful to the engineer.

The historical sketch with which the book opens proves that like other great subjects, the gas engine has long occupied men's minds.

The first six chapters treat of theory, including the distinguishing features of the gas engine method, classification, thermodynamics of the various types, and the chemical and physical phenomena of combustion and explosion.

In the seventh chapter, standard engines illustrative of the different types are described, and tests from each engine for power and consumption of gas are given. The diagrams and efficiencies are shortly discussed, compared with theory, and the various sources of loss pointed out.

The eighth chapter deals with typical igniting arrangements and the ninth with governing gear and other mechanical details.

The tenth chapter briefly describes and discusses vario[us] theories which have been propounded concerning the action [of] the gases in the cylinder of the gas engine and in gaseous expl[o]sions.

In the last chapter the great sources of loss of heat st[ill] existing in the best gas engines are discussed, with the object [of] pointing out the way still open for further advance.

Many of the tests and most of the theoretical and practic[al] discussion, result from the author's personal experience with t[he] gas engine.

In the chapter on thermodynamics the author is much indebt[ed] to the work of the late Prof. RANKINE, and he has adopted, [in] treating of efficiency, some of the elegant formulæ of Dr. AI[M]É WITZ, of Lille, to whom as well as to Prof. SCHÖTTLER and Pr[of.] THURSTON he has much pleasure in expressing his indebtednes[s.]

D. C.

BIRMINGHAM : *July* 1886.

CONTENTS.

CHAPTER		PAGE
	HISTORICAL SKETCH OF THE GAS ENGINE, 1690 TO 1885 .	1
I.	THE GAS ENGINE METHOD	23
II.	GAS ENGINES CLASSIFIED	29
III.	THERMODYNAMICS OF THE GAS ENGINE	36
IV.	THE CAUSES OF LOSS IN GAS ENGINES	72
V.	COMBUSTION AND EXPLOSION	79
VI.	EXPLOSION IN A CLOSED VESSEL	95
VII.	THE GAS ENGINES OF THE DIFFERENT TYPES IN PRACTICE	116
VIII.	IGNITING ARRANGEMENTS	202
IX.	ON SOME OTHER MECHANICAL DETAILS	226
X.	THEORIES OF THE ACTION OF THE GASES IN THE MODERN GAS ENGINE	243
XI.	THE FUTURE OF THE GAS ENGINE	260
APPENDIX		271
INDEX		275

THE GAS ENGINE.

HISTORICAL SKETCH OF THE GAS ENGINE.

THE origin of the gas engine is but imperfectly known; by some it is dated as far back as 1680, when Huyghens proposed to use gunpowder for obtaining motive power. Papin, in 1690, continued Huyghens' experiments, but without success. The method used was a fairly practicable one. The explosion was used indirectly; a small quantity of gunpowder exploded in a large cylindrical vessel filled with air, expelled the air through check valves, thus leaving, after cooling, a partial vacuum. The pressure of the atmosphere then drove a piston down to the bottom of the vessel, lifting a weight or doing other work.

In a paper, published at Leipsic in 1688, Papin stated that, 'until now all experiments have been unsuccessful; and after the combustion of the exploded powder, there always remains in the cylinder about one-fifth of its volume of air.'

The Abbé Hautefeuille made similar proposals, but does not seem to have made actual experiments. These early engines cannot be classed as gas engines. The explosion of gunpowder is so different in its nature from that of a gaseous mixture that comparison is untenable. The first real gas engine described in this country is in Robert Street's patent, No. 1983, 1794. It contains a motor cylinder in which works a piston connected to a lever, from which lever a pump is driven. The bottom of the motor cylinder is heated by a fire; a few drops of spirits of turpentine being introduced and evaporated by the heat, the motor piston is drawn up, and air entering mixes with the inflammable vapour, the application of a flame to a touch-hole causing explosion : and the piston being driven up forces the pump piston down, so performing work in raising water. The details, as described, are crude, but the main idea is correct and was not improved upon in practice till very lately.

Samuel Brown's inventions come next. His patents are dated 1823 and 1826, Nos. 4874 and 5350. The principle used is in-

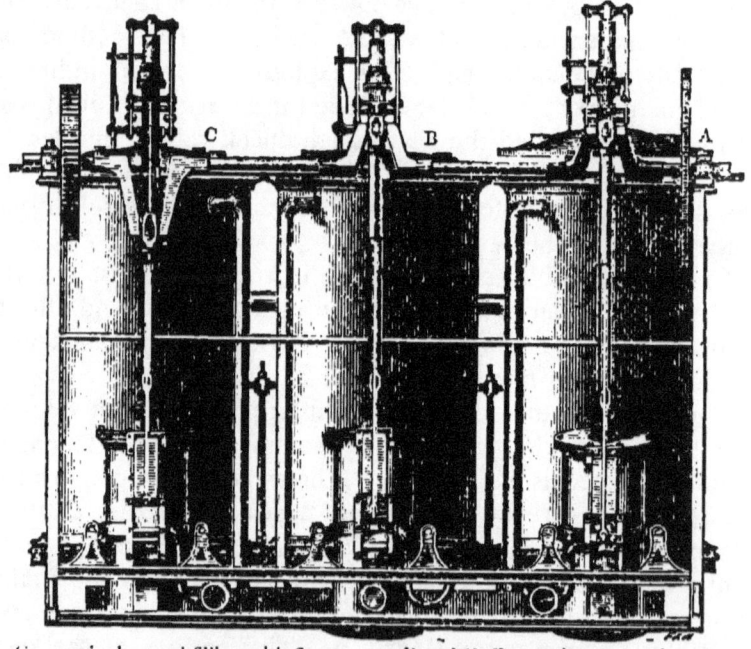

A, Cover raised, vessel filling with flame.   B and C, Covers down, vessels vacuous.
FIG. 1.—Brown's Gas-vacuum Engine, 1826.

genious, and easily carried out in practice, but it is not economical, and it gives a very cumbrous machine for the amount of power produced. A partial vacuum is produced by filling a vessel with flame, and expelling the air it contains, a jet of water is thrown in and condenses the flame, giving vacuum. The atmospheric pressure thus made available for power is utilised in any engine of ordinary construction.

Brown's apparatus consists essentially of a large upright cylindrical vessel fitted on the top with a movable valve cover, of the whole diameter of the cylinder. The cover is raised and lowered from and to its seat by a lever and suitable gear at proper times. The gas supply pipe enters the cylinder at the bottom; the cylinder being filled with air, and the valve raised, the gas cock is opened and the issuing gas lighted by a small flame as it enters the cylinder. The flame produced fills the whole vessel, expelling the air it contains; the valve being now lowered and the gas supply shut off, the water-jet is thrown in and causes condensation. To keep up a constant supply of power, several of these cylinders are required, so that one at least may be always vacuous while the others are in the process of obtaining the vacuum. In the specification three are shown and three engines. The engines are all connected to the same crank-shaft. Notwithstanding this provision, the motion must have been irregular. The idea was evidently suggested by the condensing steam engine; instead of using steam to obtain a vacuum flame is employed. Brown's engine, although uninteresting theoretically, is important as being the first gas engine undoubtedly at work. According to the 'Mechanics' Magazine,' published in London, a boat was fitted with one including a complete gas generating plant, and was run upon the Thames not for public use but only as an experiment. Another engine was made in combination with a road carriage; it also ran in London. If these statements are to be relied upon, then Samuel Brown was a really great man and should be considered as the Newcomen of the gas engine; in some points he achieved a measure of success not yet equalled by his successors.

W. L. Wright, 1833, *No.* 6525.—In this specification the drawings are very complete and the details are carefully worked out. The explosion of a mixture of inflammable gas and air acts directly upon the piston, which acts through a connecting rod upon a crank-shaft. The engine is double-acting, the piston receiving two impulses for every revolution of the crank-shaft. In appearance it resembles a high pressure steam engine of the kind known as the table pattern. The gas and air are supplied to the motor cylinder from separate pumps through two reservoirs, at a pressure a few pounds above atmosphere, the gases (gas and air) enter spherical spaces at the ends of the motor cylinder, partly displacing the previous contents, and are ignited while the piston is crossing the dead centre. The explosion pushes the piston up or down through its whole stroke; at the end of the stroke the exhaust valve opens and the products of combustion are discharged during

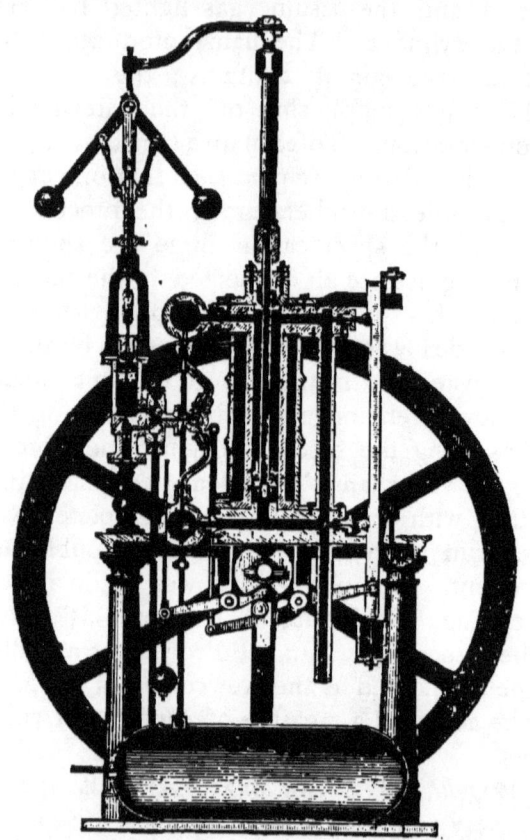

FIG. 2.—Wright's Gas-exploding Engine, 1833.

the return, excepting the portion remaining in the spaces not entered by the piston. The ignition is managed by an external flame and touch-hole. The author has been unable to find whether the engine was ever made, but the knowledge of the detail essential to a working gas engine shown by the drawings indicates that it or some similar machine had been worked by the inventor. Both cylinder and piston are water-jacketed, as would have been necessary in a double-acting gas engine to preserve the working parts from damage from the intense heat of the explosion. This is the earliest drawing in which this detail is properly shown.

William Barnett, 1838, *No.* 7615.—Barnett's inventions as described in his specification are so important that they require more complete description than has been here accorded to earlier inventors.

Barnett is the inventor of a very good form of igniting arrangement. The flame method most widely used at the present time was originated by him.

Barnett is also the inventor of the compression system now so largely used in gas engines. The Frenchman, Lebon, it is true, described an engine using compression, in the year 1801, but his cycle is not in any way similar to that proposed by Barnett, or used in the modern gas engine. Barnett describes three engines. The first is single-acting, the second and third are double-acting; all compress the explosive mixture before igniting it. In the first and second engines the inflammable gas and air is compressed by pumps into receivers separate from the motor cylinder, but communicating with it by a short port which is controlled by a piston valve. The piston valve also serves to open communication between the cylinder and the air when the motor piston discharges the exhaust gases.

In the third engine the explosive mixture is introduced into the motor cylinder by pumps, displacing as it enters the exhaust gases resulting from the previous explosion; the motor piston by its ascent or descent compresses the mixture. Part of the compression is accomplished by the charging pumps, but it is always completed in the motor cylinder itself.

In all three engines the ignition takes place when the crank is crossing the dead centre, so that the piston gets the impulse during the whole forward stroke.

Fig. 3 is a sectional elevation of the first engine, showing the principal working parts, but omitting all detail not required for explaining the action.

There are three cylinders containing pistons; A is the motor piston, B is the air pump piston. The gas pump piston cannot be seen in the section, but works in the same crosshead as B. The motor piston is suitably connected to the crank shaft, and the other two are also connected by levers in such manner that all three move simultaneously up or down. The pump pistons, moving up, take respectively air and inflammable gas into their

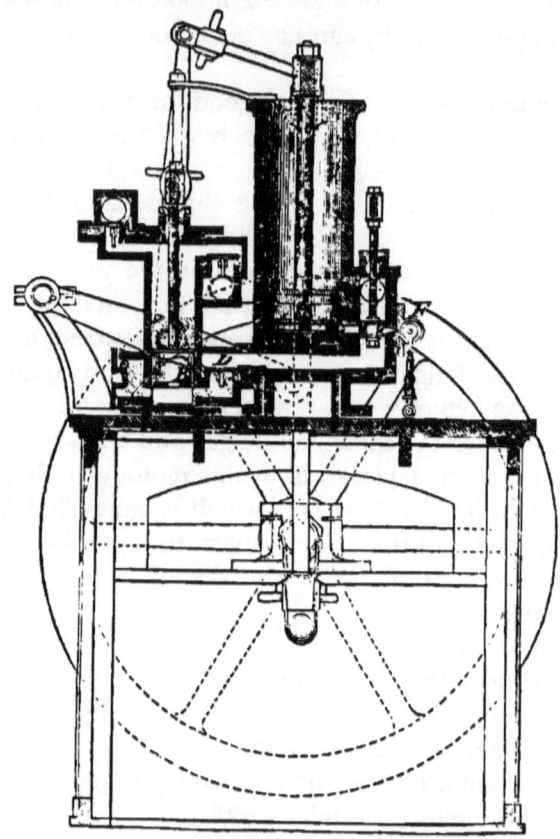

FIG. 3.—Barnett Gas Engine.

cylinders; upon the down stroke the gases are forced through an automatic lift valve into the receiver D, and there mix. When the down stroke is complete and the receiver is fully charged with the explosive mixture, the pressure has risen to about 25 lbs. per square inch above atmosphere. At the same time as the pumps are compressing, the motor piston is moving down and discharging the exhaust gases from the power cylinder; it reaches the bottom of its stroke just when compression is complete. The piston valve E then opens communication between the receiver and the motor, at the same time closing to atmosphere. The motor cylinder being in free communication with the receiver, the explosion of the mixture is accomplished by the igniting cock or valve F; the pressure resulting actuates the motor piston during its whole up-

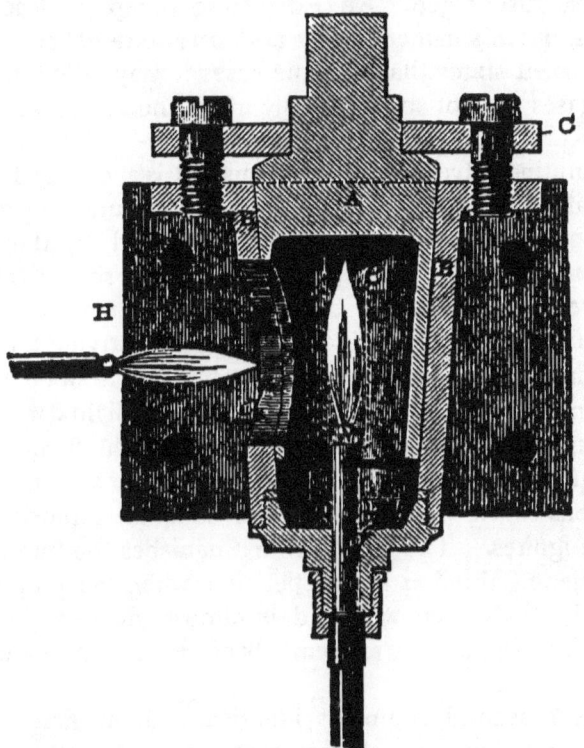

FIG. 4¹—Barnett's Igniting Cock.

ward stroke, the hot gases flowing through the port G precisely as steam would do. The volume of the receiver being constant, the pressure in the motor cylinder slowly falls by expansion, due to the movement of the piston, upon which work is performed, and by cooling, the pressure still existing in the cylinder when the stroke is complete depending on the ratio between the volume swept by the motor piston and the volume of the receiver.

The down stroke again expels the products of combustion, the valve opening to atmosphere, while the compression again takes place. This cycle gives a single-acting engine. It is obvious that as the piston A does not enter the receiver it cannot displace the exhaust gases there. If means are not taken to expel these gases they must mix with the fresh explosive charge pumped in.

It is very desirable that these gases should be as completely as possible discharged. An exhausting pump is described for doing this, but in small engines it adds an additional complication; and so Barnett states that in some cases it may be omitted. The exhaust gases do not so injuriously affect the action of small gas engines.

The igniting valve is very ingenious. It is shown at Fig. 4, on a larger scale. A hollow conical plug A is accurately ground into the shell B, and is kept in position by the gland C; the shell has two long slits D and E; the plug has one port so cut that as the plug moves it shuts to the slit D before opening to E. In the bottom of the shell there is screwed a cover carrying a gas burner F, which may be lit while the port in the plug is open to the air through D. The external constant flame H lights it. So long as the plug remains in this position the internal flame continues to burn quietly. If the plug be now turned to shut to the outer air, it opens to the slit E, and as that contains explosive mixture it at once ignites. The explosion extinguishes the internal flame, but it is again lighted at the proper time when the plug is moved round. The valve acts well and is almost identical in principle with the flame-igniting arrangements of Hugon, Otto and Langen and Otto.

Barnett's second engine is identical with his first except that it is double-acting, and therefore requires a greater number of parts.

Barnett's third engine is worthy of careful description. Fig. 5 is a vertical section of the principal parts. It is double-acting. It has three cylinders, motor, air-pump and gas-pump; the air and gas pumps are single-acting, the motor piston is double-acting. The pumps are driven from a separate shaft, which is actuated from the main crank shaft by toothed wheels; the wheel upon the pump shaft is half the diameter of that on the motor shaft, so that it makes two revolutions for one of the other. The pumps therefore make one up-and-down stroke for each up or down stroke of the motor piston; the angles of the cranks are so set that they (pumps) discharge their contents into one or other side of the motor cylinder at every stroke; the exhaust gases are partly

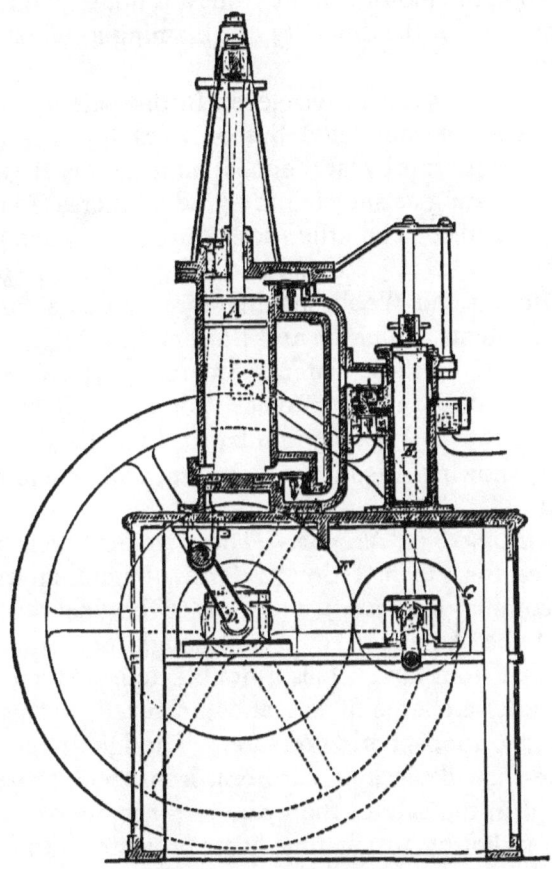

FIG. 5.—Barnett Engine.

displaced by the fresh explosive mixture, and the motor piston completes the compression in the motor cylinder itself. When full up or down the igniting cock acts, and the explosion drives the piston to the middle of its stroke; it here runs over a port in the middle of the cylinder, and the pressure at once falls to atmosphere.

A is the motor piston ; B is the air-pump piston ; C is the gas-pump piston, which is behind the air-pump, and therefore not seen in the section ; D is the main crank shaft ; E the pump shaft driven from the main shaft by the wheels F and G. The engine is exceedingly interesting as the first in which the compression is accomplished in the motor cylinder, but it is not so good a machine as the first because of the difficulty of obtaining a sufficient amount of expansion.

From 1838 to 1854 inclusive eleven British patents were applied for ; some were not completed but only reached the provisional stage. Of these patents by far the most important is Barnett's ; the others are interesting as showing the gradual increase of attention the subject attracted. The other names are Ador, 1838 ; Johnson, 1841 ; Robinson, 1843 ; Reynolds, 1844 ; Brown, 1846 ; Roger, 1853, also Bolton and Webb, making three patents for the year ; for 1854 two patents, Edington and Barsanti and Matteucci. None of the proposals in these patents are really valuable or novel, being anticipated by either Street, Wright, or Samuel Brown. Robinson's is the best, being similar to Lenoir's in some of its details, and showing distinctly a better understanding of gas engine detail.

A. V. Newton, 1855, *No.* 562.—This specification is interesting, and describes for the first time a form of igniting arrangement only now coming into use ; it seems to be identical with the invention of the American Drake, although not described as a communication from him. It is a double-acting engine, and takes into the cylinder a charge of gas and air mixed, during a portion of the stroke, at atmospheric pressure. The igniting arrangement is a thimble-shaped piece of hard cast-iron which projects into a recess formed in the side of the cylinder ; it is hollow, and is kept at all times red-hot by a blow-pipe flame projected into it by a small pump. When the piston uncovers the recess the explosive gases coming in contact with it ignite, and the pressure produced drives it forward.

This is the first instance of ignition by contact with red-hot metal ; the proposal has often been made since then in varying forms.

Barsanti and Matteucci, 1857, *No.* 1655.—This is the first free.

piston engine ever proposed; instead of allowing the explosion to act directly upon the motive power shaft through a connecting rod, at the moment of explosion the piston is perfectly free. The cylinder is very long, and is placed vertically. When the explosion occurs, it expends its power in giving the piston velocity; the expansion therefore takes place with considerable rapidity, and the piston, gaining speed until the pressure upon it falls to atmosphere, moves on, till the energy of motion is absorbed doing work on the external air, lifting the piston and in friction. When the energy is all absorbed in this manner it stops; it has reached the top of its stroke. A partial vacuum has been formed in the cylinder, and the weight has been raised through the stroke. It now returns under the pressure of the atmosphere and its own weight; in returning, a rack attached to the piston engages the motive shaft and drives it. The cooling of the gases as the piston descends continues and helps to keep up the vacuum.

The method although indirect is economical. Three advantages are gained by it—rapid expansion, considerable expansion (an expansion of six times is common in these engines), and also some of the advantages of a condensor.

Fig. 6 shows a vertical section of their best modification. The motor piston A working in the tall vertical cylinder B is attached to the rack C, which works into the toothed wheel D. The motor shaft E revolves in the direction of the arrow, and it is provided with a ratchet; a pall upon the wheel D engages the ratchet on the down stroke of the piston only, on the up stroke it slips freely past the ratchet. The piston A is therefore quite free to move without the shaft on the up stroke, but it engages on the down stroke. The cams F and G are arranged to strike projections upon the rack, and so raise or lower the piston. It is raised when the charge is to be taken in, and lowered when it has completed its working stroke and the exhaust gases have to be discharged. When raised the valve H is in the position shown. Air first enters the cylinder through the port I, which also serves to discharge the exhaust. After the piston has uncovered the port K the valve H shuts on I, opening at the same time on K; the gas supply then enters and mixes more or less perfectly with the air previously introduced.

A small further movement of the piston now closes the valve, and the explosion is caused by the passage of the electric spark in the position indicated upon the drawing. The piston shoots up

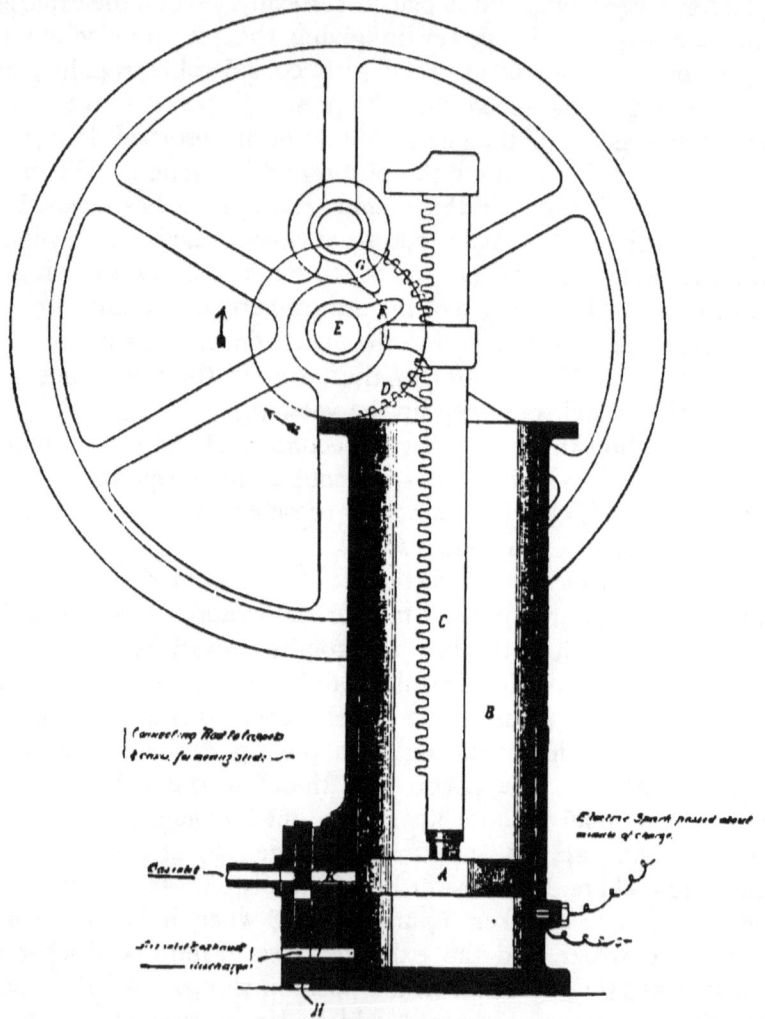

FIG. 6.—Barsanti and Matteucci Engine, 1857.

freely to the top of its stroke, to give out the work stored up usefully upon its return.

As the next engine to be described marks the beginning of the practicable stage of gas engine development, it is advisable to summarise before proceeding.

Previous to 1860 the gas engine was entirely in the experimental stage. Many attempts were made, but none of the inventors sufficiently overcame the practical difficulties to make any of their engines commercially successful. This was mostly due to the very serious nature of the difficulties themselves, but it was also due to too great ambition of the inventors; they wished not only to compete with the steam engine for small powers, but for large powers. They thought in fact more to displace the steam engine than to compete with it.

This is clearly shown in many of their descriptions of the applications of their inventions.

The greatest credit is due to Wright and Barnett. Wright very closely proposed the modern non-compression system, Barnett the modern compression system. Barnett is also the originator of one of the modern flame systems for ignition. Barsanti and Matteucci follow in order of merit as the inventors of the free-piston gas engine.

M. Lenoir occupies the honourable position of the inventor of the first gas engine ever actually introduced to public use. The engine was not strikingly novel; nothing was done in it which had not been proposed before, but its details were thoroughly and carefully worked out. It was in fact the first to emerge from the purely experimental stage. Lenoir's real credit consists in overcoming the practical difficulties sufficiently to make previous proposals fairly workable.

The principle is exceedingly simple and evident. The piston moves forward for a portion of its stroke, by the energy stored in the fly wheel, and takes into the cylinder a charge of gas and air at the ordinary atmospheric pressure. The valves cut off communication, and the explosion is occasioned by the electric spark; this propels the piston to the end of the stroke. Exhausting is done precisely as in the steam engine.

The engine is simply an ordinary high-pressure steam engine with valves arranged to admit gas and air and discharge the products of combustion. Fig. 7 is an external elevation of a three-horse engine. It was first constructed in Paris in 1860 by M. Hippolyte

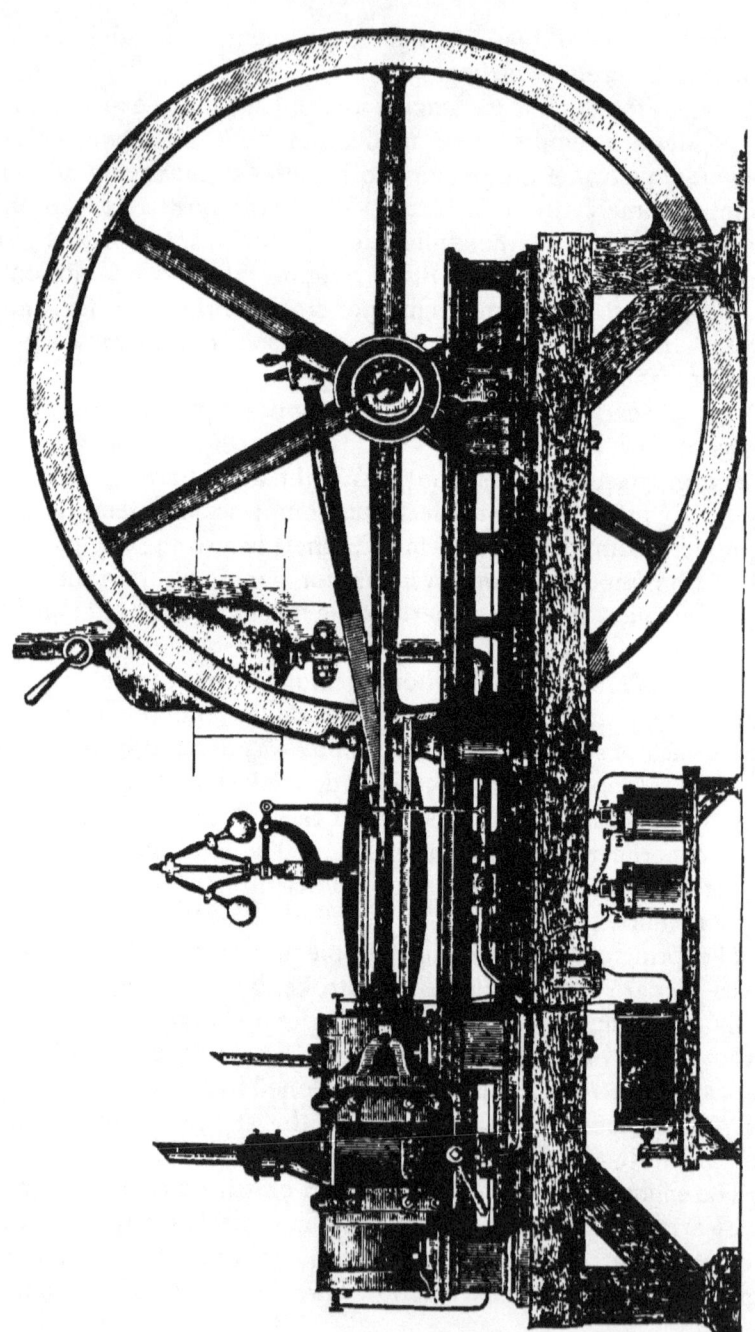

FIG. 7.—Lenoir's Gas Engine.

Marinoni. In Moigno's 'Cosmos' of that year it is stated that two engines were in course of manufacture, one of six horse-power, the other of twenty.

The early statements of its economy were ludicrously inaccurate. A one horse-power engine consumed, it was said, but 3 cubic metres (106 cubic ft. nearly) of coal gas in twelve hours' work, and therefore cost for fuel not more than one-half of what a steam engine would have done.

The actual consumption was speedily shown to be much nearer 3 cubic metres per effective horse-power per hour.

Notwithstanding the high consumption, the engine had many good points; its action was exceedingly smooth; no shock whatever was heard from the explosion. Indeed it is quite impossible when watching the engine in motion to realise that regular explosions are occurring. The motion is as smooth and silent as in the best steam engines.

In the 'Practical Mechanics' Journal' of August 1865, there is an article describing the progress made by the engine since the date of its introduction, from which it appears that in Paris and France from 300 to 400 engines were then at work, the power ranging from half horse to three horse.

The Reading Iron Works Company, Limited, at Reading, undertook the manufacture for this country. One hundred engines were made and delivered by them; several of them have continued at work till now. Notably one engine inspected by the author at Petworth House, Petworth, worked for twenty years pumping water, and is even yet in good condition.

The work performed by the engines was multifarious in its character—printing, pumping water, driving lathes, cutting chaff, sawing stone, polishing marble, in fact, wherever from one-half to three horse-power was sufficient.

Lenoir's patent in this country was obtained by J. H. Johnson, 1860, No. 335. It describes very closely the engine as manufactured both in France and England. The subsequent patent, 1861, No. 107, does not seem to have been carried into effect.

These specifications contain many erroneous ideas, showing the notions then prevalent among inventors of the nature of gaseous explosions. Lenoir erroneously supposed that the economy of his engine would be improved if he could obtain a slower explosion. He evidently thought that the power imparted to the piston by explosion was similar in nature to a sudden blow—a rapid rise of pressure, and a fall nearly as rapid. He therefore attempted to avoid explosion by such expedients as stratification and injection of steam or water spray. The stratification idea he very clearly expressed in his second specification, stating that 'the object of preventing the admixture of air and gas is to avoid explosion.' It is somewhat extraordinary to find notions so erroneous common at a time when Bunsen's work had clearly proved the continuous nature of the combustion in gaseous explosions, and when Hirn had made experiments which showed that the heat evolved by explosion in a gas engine was only a small part of the total heat of the combustion, the heat which did not appear during explosion being produced during expansion.

Other speculations on the cause of the uneconomical working of the engine were frequent, but the true reason was fully explained by Gustav Schmidt in a paper read before 'The Society of German Engineers' in 1861. He states : 'The results would be far more favourable if compression pumps, worked from the engine, compressed the cold air and cold gas to three atmospheres before entrance into the cylinder ; by this a greater expansion and transformation of heat is possible.'

This opinion became common at this time. Compression engines were proposed with great clearness and a full understanding of the advantages to be gained.

Million, 1861, *No.* 1840.—This Frenchman had exceedingly clear ideas of the advantages of compression ; he evidently considers himself as the first to propose its use in a gas engine, apparently unaware of the existence of Barnett's engine already described. He claims the exclusive right to use compression in the most emphatic language.

The first engine described is exactly what Schmidt asks for. Separate pumps compress the air and gas into a reservoir, from which the movement of the motor piston, during a portion of the stroke, withdraws its charge under compression. Ignition is accomplished by the electric spark, and the piston moves forward under the high pressure produced. He states :

'In ordinary air engines the operation of the motive cylinders is analogous to that of the pumps, the result being that there are two cylinders, which act in directions contrary to each other, and that the pump, which is an organ of resistance, even works at a greater pressure than that of the motive cylinder, which is an organ of power. Thus these engines are very large in proportion to their power. On the contrary by employing gases under the conditions above explained, these engines will exert great power in proportion to their dimensions. The sudden ignition of the gases in the motive cylinder causes the latter to work at an operative pressure much greater than that of the pumps.'

The advantage of compression in a gas engine could not be more fully and clearly stated. But he goes even a step further; he sees that the portion of the motor piston stroke spent in taking in the charge under compression, is a disadvantage, and he proposes to make the whole stroke available for power by providing a space at the end of the cylinder in which the gases are compressed.

'Instead of introducing the cold gases into the cylinders, during a portion of the stroke and igniting them afterwards, when the induction ceases . . . another arrangement might be adopted. The motive cylinder might be made longer than necessary, in order that the piston should always leave between it and the end of the cylinder a greater or less space, according to the pleasure of the constructor, such as one-fourth or one-third, more or less, of the volume generated by the motive piston. This space is called by the inventor a cartridge. On opening the slide valve the gases could be allowed to enter suddenly from the pressure reservoir into this cartridge towards the dead point, and this induction having ceased, an electric spark would ignite the gases in the cartridge by which the driving piston would be set in motion.'

Such an engine would resemble in its action the best modern compression engines. The difficulties of ignition however are too considerable to be overcome without further detail.

The compression idea at this date was evidently widely spread, because it again crops up in a remarkably clever pamphlet by M. Alph. Beau de Rochas, published at Paris in 1862. He advances a step further than Million, and investigates the conditions of greatest economy in gas engines using compression, with reference to volume of hot gases and surfaces exposed. He states that to obtain economy with an explosion engine, four conditions are requisite :

1. The greatest possible cylinder volume with the least possible cooling surface.
2. The greatest possible rapidity of expansion.
3. The greatest possible expansion ; and
4. The greatest possible pressure at the commencement of the expansion.

In using boiler tubes, he states, the efficiency of the heat transmitted increases with reduction in the diameter of the tubes. In the case of engine cylinders, therefore, the loss of heat of explosion would be in inverse ratio to the diameter of the cylinders.

Therefore, he reasons, an arrangement which for a given consumption of gas, gives cylinders of the greatest diameters, will give the best economy, or least loss of heat to the cylinder. One cylinder only must be employed in such an engine.

But loss of heat depends also upon time ; cooling, therefore, will be proportionately greater as the working speed is slower.

The sole arrangement capable of combining these conditions, he states, consists in using the largest possible cylinder, and reducing the resistance of the gases to a minimum. This leads, he states, to the following series of operations.

1. Suction during an entire outstroke of the piston.
2. Compression during the following instroke.
3. Ignition at the dead point and expansion during the third stroke.
4. Forcing out of the burned gases from the cylinder on the fourth and last return stroke.

The ignition he proposes to accomplish by the increase of temperature due to compression. This he expects to do by compressing to one-fourth of the original volume.

In our own country the late Sir C. W. Siemens proposed compression in 1862. The idea was exceedingly widely spread, as is evident from those numerous and independent inventions. The practical experience to enable it to be successfully effected had yet to be created, however, and this took many years of patient work.

Historical Sketch of the Gas Engine

The igniting arrangement was the first weak point requiring improvement. The electrical method of Lenoir was exceedingly delicate and troublesome.

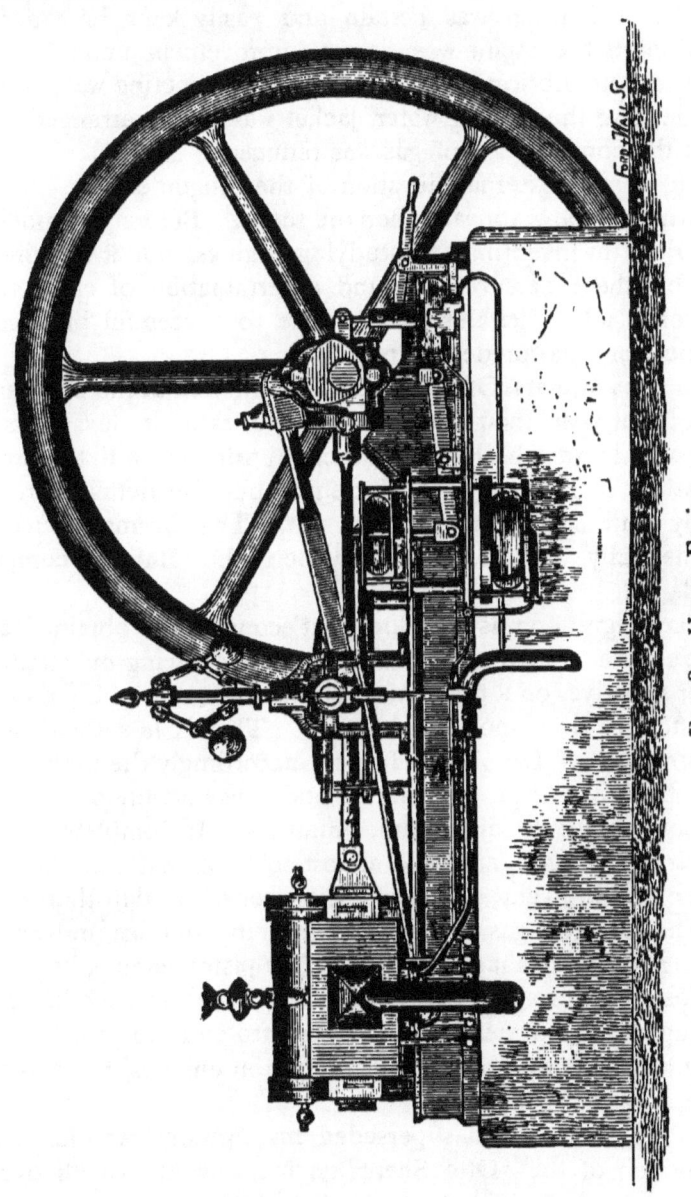

FIG. 8 – Hugon Engine.

Hugon's engine, produced in 1865, was similar to Lenoir's; but the igniting was accomplished by flame, a modification of Barnett's, 1838, using a slide valve instead of a lighting cock. The flame ignition was certain and easily kept in order. In other points the engine was a great improvement upon its predecessor. The lubrication was improved by injecting water into the cylinder and the cooling water jacket was better arranged. As a result the consumption of gas was reduced.

Fig. 8 is an external elevation of the Hugon engine.

Mr. Otto now appears upon the scene. Before him much had been done in inventing and studying engines, but it remained for him by sheer perseverance and determination of character, to overcome all difficulties and reduce to successful practice the theories of his predecessors.

In 1867 Messrs. Otto and Langen exhibited at the Paris exhibition of that year, their free piston engine, exterior elevation shown at fig. 9. It was absolutely identical in principle with the previous invention of Barsanti and Matteucci, but the details were completely and successfully carried out. The Germans succeeded commercially and scientifically when the Italians completely failed.

Flame ignition was used and great economy was obtained, a half-horse engine, according to Professor Tresca, giving over half-horse power effective, on a gas consumption at the rate of 44 cubic feet per effective horse-power per hour. This is less than half the consumption of Lenoir or Hugon; accordingly the prejudice excited by the strange appearance and noisy action of the engine did not prevent its sale in large numbers. It completely crushed Lenoir and Hugon, and held almost sole command of the market for ten years, several thousands being constructed in that period.

The Brayton gas engine appeared in America in 1873, but although more mechanical than any free piston engine, its economy was insufficient to enable it to compete. It was better than Lenoir or Hugon, but not nearly so good as Otto and Langen.

Other inventors attempted free piston engines, but with small success.

In 1876 Mr. Otto superseded his former invention by the production of the 'Otto Silent' engine, now known all over the globe. It is a compression engine, using the precise cycle described in 1862 by Beau de Rochas, but carried out in a most perfect manner and using a good form of flame ignition, a modified Otto and Langen valve in fact. The economy is greater than that of any

FIG. 9.—Otto and Langen Free Piston Engine.

previous engine, one indicated horse being obtained upon 20 cubic feet of gas, or one effective horse upon 24 to 30 cubic feet per hour.

This engine has established gas engines upon a firm commercial basis, 15,000 having been sold since its invention ; this represents at least an effective power of 90,000 horses.

Strangely enough, although Mr. Otto is the greatest and most successful gas engine inventor who has yet appeared, he adheres to Lenoir's erroneous ideas, and in his specification 2081 of 1876 he attributes the economy of his machine to a slow explosion caused by arrangement of gases within the cylinder.

The compression, which is the real cause of the economy and efficiency of the machine, he seems to consider as an accidental and unessential feature of his invention.

The gas engine, like all great inventions, is the result of the long-continued labour of many minds; it is a gradual growth due to the united labours of many inventors. In the earlier days of motive power, explosion was as much in the minds of the inventors, Huyghens and Papin, as steam, but the mechanical difficulties proved too great. The constructive skill of the time was heavily taxed by the rude steam engine of Newcomen, and still more unequal to the invention of James Watt; it was in 1774 that Watt ran his first successful steam engine at Soho Works, Birmingham. Twenty years later, 1794, Street's gas engine patent indicated the direction of men's minds, seeking a rival for steam before steam had been completely introduced. The experience and skill accumulating in the construction of the steam engine made the gas engine more and more possible.

The proposals of Brown, 1823; Wright, 1833; Barnett, 1838; Barsanti and Matteucci, 1857, show gradually increasing knowledge of detail and the difficulties to be overcome, all leading to the first practicable engine in 1860, the Lenoir.

Since that date till now, twenty-five years, great advances have been made, and at present the gas engine is the only real rival to steam.

CHAPTER I.

THE GAS ENGINE METHOD.

GAS ENGINES, while differing widely in theory of action and mechanical construction, possess one feature in common which distinguishes them from other heat engines : that feature is the method of heating the working fluid.

The working fluid is atmospheric air, and the fuel required to heat it is inflammable gas. In all gas engines yet produced, the air and gas are mixed intimately with each other before introduction to the motive cylinder ; that is, the working fluid and the fuel to supply it with heat are mixed with each other before the combustion of the fuel.

The fuel, which, in the steam and in most hot-air engines, is burned in a separate furnace, is, in the gas engine, introduced directly to the motive cylinder and burned there. It is indeed part of the working fluid.

This method of heating may be called the gas-engine method, and from it arises at once the great advantages and also the great difficulties of these motors.

Compare first with the steam engine. In it there exist two great causes of loss : water is converted into steam, absorbing a great amount of heat in passing from the liquid to the gaseous state ; after it has been used in the engine it is rejected into the atmosphere or the condenser, still existing as steam. The heat necessary to convert it from the liquid to the gas is consequently in most part rejected with it. Loss, occurring in this way, would be small if high temperatures could be used ; but this is the point where steam fails. High temperatures cannot be obtained without pressure so great as to be quite unmanageable. The attempt to obtain high temperatures by super-heating has often been made, but without any substantial success. Although the difficulty of excessive pressure is avoided, another set of troubles are introduced. All the heat to be given to the gaseous steam must pass through the iron plates forming the boiler or super-heater, which plates will only stand a comparatively low temperature, certainly not exceeding that of a low red heat, or about 600° to 700° C. Steam, being a gas, is much more difficult to heat than water; it follows that even these temperatures cannot be attained without enormous addition to the heating surface. The difficulties of making a workable engine using high temperature steam are so great that even so distinguished an engineer and physicist as the late Sir C. W. Siemens failed in his attempts, which extended over many years. It may be taken then that low temperature is the natural and unavoidable accompaniment of the steam method, arising from the necessary change of the physical state of the working fluid, and the limited temperature which iron will safely bear. The originators of the science of thermodynamics have long taught that the maximum efficiency of a heat engine is obtained when there is the maximum difference between the highest and lowest temperatures of the working fluid. So long ago as 1854, Professor Rankine read a paper before the British Association, 'On the means of realising the advantages of the Air Engine,' in which he expresses his belief that such engines will be found to be the most economical means of developing motive power by the agency of heat. In this opinion he stood by no means alone. Engineers so able as Stirling, Ericsson, and Siemens; physicists so distinguished as Dr. Joule, and Sir Wm. Thomson, devoted much energy and study to their practice and theory. Notwithstanding all their efforts, aided by a host of less able inventors, the difficulties proved too formidable; and although more than thirty years have now passed since Rankine announced his belief, the hot-air engine proper, has made no real advance. Similar causes to those acting in the steam engine impose a limit here. It is true the complication of changing physical state is avoided, but the limited resistance of iron to heat acts as powerfully as ever. Air is much more difficult to heat than water, and, therefore, requires a much larger surface per unit of heat absorbed. In the larger hot-air engines, accordingly, the furnaces and heating surfaces gave great trouble. Very low maximum temperatures were attained in practice. In a Stirling engine giving out thirty-seven brake horse-power, the maximum temperature was only 343° C. ; in the engines of the ship 'Ericsson,' the maximum was only about 212° C., according to Rankine, the indicated power being about 300 horses. These figures show that the heating surfaces were insufficient, as in both cases the furnaces were pushed to heat the metal to a good red. A method of internal firing was proposed, first by Sir George Cayley and afterwards carried out with some success by others ; the furnace was contained in a completely closed vessel, and the air to be heated was forced through it before passing to the motor cylinder. The plan gave better results, but the temperature of 700° C. was still the limit, as the strength of the iron reservoir had to be considered, and the hot gases had to pass through valves. Wenham's engine, described in a paper read before the Institution of Mechanical Engineers in 1873, is a good example of this class. In it the highest temperature of the working fluid, as measured by a pyrometer, was 608° C. ; higher temperatures could easily have been got but the safety of the engine did not permit it. Professor Rankine in his work on the steam engine has very fully discussed the disadvantages arising from low maximum temperatures. He calculates that in a perfect air engine without regenerator an average pressure of 8·3 lbs. per square inch would only be attained with a maximum of 216·6 lbs. per square inch, thus necessitating great strength of cylinder and working parts for a very small return in effective power. In the 'Ericsson,' the average effective pressure was less than this, being only about 2 lbs. per square inch ; it had four air cylinders each of 14 feet diameter, and only indicated 300 horse-power. Stirling's motor cylinder did not give a true idea of the bulk of the engine, as the real air-displacer was separate. Even with Wenham's machine the bulk was excessive, an engine of 24 inches diameter cylinder and 12 inches stroke giving 4 horse-power.

Those facts sufficiently illustrate the practical difficulties which prevented the development of the hot-air engine proper. All flow from the method of heating. Low temperature is necessary to secure durability of the iron.

All hot-air engines are, therefore, very large and very heavy for the power they are capable of exerting.

The friction of the parts is so great that although the theoretical efficiency of the working fluid is higher than in the best steam engines, the practical efficiency or result per horse available for external work is not nearly so great. The best result ever claimed for Stirling's engine is 2·7 lbs. of coal per bk. horse-power per hour, probably under the truth, but even allowing it, a first class steam engine of to-day will do much better. According to Prof. Norton, the engines of the 'Ericsson' used 1·87 lbs. of anthracite per indicated horse-power per hour; but the friction must have been enormous. Compared with the steam engine, the practical disadvantages of the hot-air engine are much greater than its advantage of theory. Owing to the great inferiority of air to boiling water as a medium for the convection of heat, the efficiency of the furnace is much lower; owing to the high maximum and low available pressure, the friction is much greater—which disadvantages in practice more than extinguish the higher theoretical efficiency.

The gas engine method of heating by combustion or explosion at once disposes of those troubles; it not only widens the limits of the temperatures at command almost indefinitely, but the causes of failure with the old method become the very causes of success with the new method.

The difficulty of heating even the greatest masses of air is quite abolished. The rapidly moving flash of chemical action makes it easy to heat any mass, however great, in a minute fraction of a second; when once heated the comparatively gradual convection makes the cooling a very slow matter. The conductivity of air for heat is but slight, and both losing and receiving heat from enclosing walls are carried on by the process of convection, the larger the mass of air the smaller the cooling surface relatively. Therefore the larger the volumes of air used, the more economical the new method, the more difficult the old. The low conductivity for heat, the cause of great trouble in hot-air machines, becomes the unexpected cause of economy in gas engines. If air were a rapid carrier of heat, cold cylinder gas engines would be impossible. The loss to the sides of the enclosing cylinders would be so great that but little useful effect could be obtained. Even as it is, present loss from this cause is sufficiently heavy. In the earlier engines as much as three-fourths of the whole heat of the combustion was lost in this way; in the best modern engines so much as one-half is still lost.

A little consideration of what is occurring in the gas engine cylinder at each explosion will show that this is not surprising. Platinum, the most infusible of metals, melts at about 1700° C.; the ordinary temperature of cast iron flowing from a cupola is about 1200° C.; a temperature very usual in a gas engine cylinder is 1600° C., a dazzling white-heat. The whole of the gases filling the cylinder are at this high temperature. If one could see the interior it would appear to be filled with a blinding glare of light. This experiment the writer has tried by means of a small aperture covered with a heavy glass plate, carefully protected from the heat of the explosion by a long cold tube. On looking through this window while the engine is at work, a continuous glare of white light is observed. A look into the interior of a boiler furnace gives a good notion of the flame filling the cylinder of a gas engine.

At first sight it seems strange that such temperature can be used with impunity in a working cylinder; here the convenience of the method becomes evident. The heating being quite independent of the temperature of the walls of the cylinder, by the use of a water-jacket they can be kept at any desired temperature. The same property of rapid convection of heat, so useful for generating steam from water, is essential in the gas engine to keep the rubbing surfaces at a reasonable working temperature. In this there is no difficulty, and notwithstanding the high temperature of the gases, the metal itself never exceeds the boiling point of water.

So good a result cannot of course be obtained without careful proportioning of the cooling surfaces for the amount of heat to be carried away; in all modern engines this is carefully attended to, with the gratifying result that the cylinders take and retain a polished surface for years of work just as in a good steam engine.

The gas engine method gives the advantage of higher temperature of working fluid than is attainable in any other heat engine, at the same time the working cylinder metal may be kept as cool as in the steam engine. It also allows of any desired rate of heating the working fluid in any required volumes.

In consequence of high temperatures the available pressures are high, and therefore the bulk of the engine is small for the power obtained.

It realises all the thermodynamic advantages claimed for the hot-air engine without sacrificing the high available pressures and rapid rate of the generation of power which is the characteristic of the steam engine.

For rapid convection of heat existing in the steam boiler is substituted the still more rapid heating by explosion or combustion, a rapidity so superior that the power is generated for each stroke separately as required, there being no necessity to collect a great magazine of energy.

The only item to the debtor side of the gas engine account is the flow of heat through the cylinder walls, which disadvantage is far more than paid for by the advantages.

CHAPTER II.

GAS ENGINES CLASSIFIED.

ALTHOUGH the gas engine patents now in existence number many hundreds, the essential differences between the inventions are not great. In their working process they may be divided into a few well-defined types :

1. Engines igniting at constant volume, but without previous compression.
2. Engines igniting at constant pressure, with previous compression.
3. Engines igniting at constant volume, with previous compression.

THE FIRST TYPE is the simplest in idea ; it is the most apparent method of obtaining power from an explosion.

In it the engine draws into its cylinder gas and air at atmospheric pressure, for a part of its stroke, in proportions suitable for explosion ; then a valve closes the cylinder, and the mixture is ignited. The pressure produced pushes forward the piston for the remainder of its travel, and upon the return stroke the products of the combustion are expelled exactly as the exhaust of a steam engine. By repeating the same process on the other side of the piston, a kind of double-acting engine is obtained. It is not truly double-acting, as the motive impulse is not applied during the whole stroke, but only during that portion of it left free after performing the necessary function of charging with the explosive mixture.

The working cycle of the engine consists of four operations :

1. Charging the cylinder with explosive mixture.
2. Exploding the charge.
3. Expanding after explosion.
4. Expelling the burned gases.

To carry it out in a perfect manner, the mechanism must be so arranged that during the charging, the pressure of the gases in the cylinder does not fall below atmosphere; there must be no throttling of the entering gases. The cut-off and the explosion must be absolutely simultaneous and also instantaneous, so that the heat may be applied without change of volume, and thereby produce the highest pressure which the mixture used is capable of giving. The expansion will be carried far enough to reduce the pressure of the explosion to atmosphere; and the exhaust stroke will be accomplished without back pressure. The charge in entering must not be heated by the walls of the cylinder, but should remain at the temperature of the atmosphere till the very moment previous to ignition. At the same time, the cylinder should not cool the gases after the explosion, no heat should disappear except through expansion doing work.

Although all these conditions are necessary to the perfect cycle, it is evident that no actual engine is capable of combining them. Some throttling at the admission of the mixture, and a little back pressure during the exhausting are unavoidable; some time must elapse between the closing of the inlet valve and the explosion, in addition to the time taken by the explosion itself. Heat will be communicated to the entering gases and lost by the exploded gases to the walls of the cylinder.

The actual diagram taken from an engine will therefore differ considerably from the theoretic one.

The theoretical conditions are to a great extent contradictory.

The idea of the type, however, is easily comprehensible, and evidently suggested by the common knowledge of the destructive effect of accidental coal gas explosions which occurred soon after the introduction of gas into general use. 'The power is there, let us use it like steam in the cylinder of a steam engine,' said the early inventors.

The two most successful engines of this type were Lenoir's and, later, Hugon's, for very small powers ranging from one man to half-horse. Simple forms of this type are still in extensive use. The most widely known of these is the Bisschof, a French invention.

THE SECOND TYPE is not so simple in its main idea, and required much greater knowledge of detail, both mechanical and theoretical. As a hot-air engine its theory was originally proposed by Sir Geo. Cayley, and, later, by Dr. Joule and Sir Wm. Thomson. As a hot-air engine it failed for the reasons discussed in the previous chapter.

In it the engine is provided with two cylinders of unequal capacity; the smaller serves as a pump for receiving the charge and compressing it, the larger is the motor cylinder, in which the charge is expanded during ignition and subsequent to it.

The pump piston, in moving forward, takes in the charge at atmospheric pressure, in returning compresses it into an intermediate receiver, from which it passes into the motor cylinder in a compressed state. A contrivance similar to the wire gauze in a Davy lamp commands the passage between the receiver and the cylinder, and permits the mixture to be ignited on the cylinder side as it flows in without the flame passing back into the receiver.

The motor cylinder thus receives its working fluid in the state of flame, at a pressure equal to, but never greater than, the pressure of compression. At the proper time, the valve between the motor and the receiver is shut, and the piston expands the ignited gases till it reaches the end of its stroke, when the exhaust valve is opened, and the return expels the burned gases.

The ignition here does not increase the pressure, but increases the volume. The pump, say, puts one volume or cubic foot into the receiver; the flame causes it to expand while entering the cylinder to two cubic feet. It does the work of two cubic feet in the motor cylinder, so that, though there is no increase of pressure, there is nevertheless an excess of power over that spent in compressing.

In the first type of engine the heat is given to the working fluid at constant volume, in the second type the heat is given to the working fluid at constant pressure during change of volume.

The working cycle of the engine consists of five operations:
1. Charging the pump cylinder with gas and air mixture.
2. Compressing the charge into an intermediate receiver.
3. Admitting the charge to the motor cylinder in the state of flame, at the pressure of compression.

4. Expanding after admission.

5. Expelling the burned gases.

To carry out the process perfectly the following conditions would be required.

No throttling during admission of the charge to the pump.

No heating of the charge as it enters the pump from the atmosphere.

No loss of the heat of compression to the pump and receiver walls.

No throttling as the charge enters the motor cylinder from the receiver.

No loss of heat by the flame to the sides of the motor cylinder and piston.

And last, No back pressure during the exhaust stroke.

The exhaust gases also must be completely expelled by the motor piston; that is, the motor cylinder should have no clearance.

The requirements of this type, although sufficiently numerous and exacting, are not so contradictory among themselves as in the first.

Although every engine of the kind yet made fails to fulfil them, it is quite possible that a machine very closely approximating may be yet constructed.

The most successful engines of this kind have been Brayton's and Simon's, the first an American invention, and the second an English adaptation of it. Sir C. W. Siemens proposed such an engine in 1861, but does not seem to have been successful in carrying it out. In 1860 it was also proposed by F. Million, but without a sufficient understanding of the mechanical detail necessary for a working machine.

Brayton's engine was made in considerable numbers in America, and was applied by him to drive a good-sized launch, petroleum being used as the fuel instead of gas. It was exhibited at the Centennial Exhibition in Philadelphia; at the Paris exhibition of 1878 by Simon.

THE THIRD TYPE is the best kind of compression engine yet introduced; by far the largest number of gas engines in every day use throughout the world are made in accordance with its requirements. In theory it is more easily understood as requiring two cylinders, compression and power.

The leading idea, compression and ignition at constant volume, was first proposed by Barnett in 1838, then by Schmidt in more general terms, very fully by Beau de Rochas in 1860 and also by F. Million in the same year. Otto, however, was the first to successfully apply it, which he did in 1876.

The compression cylinder may be supposed to take in the charge of gas and air at atmospheric temperature and pressure; compress it into a receiver from which the motor cylinder is supplied; the motor piston to take in its charge from the reservoir in a compressed state; and then communication to be cut off and the compressed charge ignited.

Here ignition is supposed to occur at constant volume, that is, the whole volume of mixture is first introduced and then fired; the pressure therefore increases. The power is obtained by igniting while the volume remains stationary and the pressure increases.

Under the pressure so produced, the piston completes its stroke, and upon the return stroke the products of the combustion are expelled.

In this case the working cycle of the engine consists of six operations:

1. Charging the pump cylinder with gas and air mixture.
2. Compressing the charge into an intermediate receiver.
3. Admitting the charge to the motor cylinder under compression.
4. Igniting the mixture after admission to the motor.
5. Expanding the hot gases after ignition.
6. Expelling the burned gases.

To carry out the process perfectly, similar conditions are necessary to those in the second type. But the conditions are more contradictory. The gases entering the cylinder under pressure must not be heated by its walls; no heat should be added till the ignition; then, after ignition the gases must not lose heat to the cylinder—conditions which it is impossible for the same cylinder to fulfil simultaneously.

In the engines constructed the receiver is dispensed with, for reasons which will be explained in discussing the practical difficulties of construction ; but this does not in any way modify the theory, which shall first be discussed.

The most considerably used engine of this kind is the Otto, next to it coming Clerk's engine, then Robson's by the Messrs. Tangye, and Andrews' Stockport compression engine. In none of these types does any part of the working cycle require either the heating or the cooling of the working fluid by the relatively slow processes of convection and conduction.

Heating is accomplished by the rapid method of explosion or, if the term be preferred, combustion, and for the cooling necessary in all heat engines is substituted the complete rejection of the working fluid with the heat it contains and its replacement by a fresh portion taken from the atmosphere at the atmospheric temperature, which is the lower limit of the engines.

This is the reason why those cycles can be repeated with almost indefinite rapidity, and why gas engines can be run at speeds equal to steam engines, while the old hot-air engines could not be run fast, because of the very slow rate at which air could be heated and cooled by contact.

There still remains one important type of gas engine not included in this classification ; in it part of the efficiency is dependent on cooling by contact, and consequently only a slow rate of working stroke can be obtained. It is the kind of engine known as the free piston or atmospheric gas engine. It may be regarded as a modification of the first type. The first part of its action is precisely similar, the latter part differs considerably.

It may be called *Type* ONE A. In it the piston moves forward, taking in its charge of gas and air from the atmosphere at the atmospheric pressure and temperature. When cut off it is ignited instantaneously, the volume being constant and the pressure increasing ; the piston is not connected directly to the motor shaft, but is free to move under the pressure of the explosion, like the ball in a cannon. It is shot forward in the cylinder (which is made purposely very long) ; the energy of the explosion gives the piston velocity ; it therefore continues to move considerably after the pressure has fallen by expansion to atmosphere ; a partial vacuum forms under the piston till its whole energy of motion is absorbed in doing work upon the exterior air. It then stops, and the external pressure causes it to perform its instroke, during which a clutch arrangement yokes it to the motor shaft, giving the shaft an impulse. The explosion is made to give its equivalent in work upon the external air, in forming a vacuum in fact; the vacuum is increased by the cooling of the hot gases during the return of the piston. The piston proceeds completely to the bottom of the cylinder, expelling the products of combustion. So far as the working fluid of the engine is concerned the cycle consists of five operations:

1. Charging the cylinder with explosive mixture.
2. Exploding the charge.
3. Expanding after explosion.
4. Compressing the burned gases after some cooling.
5. Expelling the burned gases.

To carry it out perfectly, in addition to the requirements of the first type, the expansion should be carried far enough to lower the temperature of the working fluid to the temperature of the atmosphere, and the compression to atmospheric pressure again should be conducted at that temperature; that is, the compression line should be an isothermal.

This kind of engine was proposed first by Barsanti and Matteucci in 1854, by F. H. Wenham in 1864, and then by Otto and Langen in 1866. The last named inventors were successful in overcoming the practical difficulties, and many engines were made and sold by them. Their engine, although cumbrous and noisy, was a good and economical worker; many are still in use. The next best known engine of the kind was Gillies's, of which a considerable number were constructed and sold.

CHAPTER III.

THERMODYNAMICS OF THE GAS ENGINE.

BEGINNING with Professor Rankine, able writers have so fully treated the thermodynamics of the air engine that but little can be added to the knowledge of the subject now in existence. The gas engine method of heating, however, introduces limits of temperature so extended and cycles of action so different from those possible in the air engine proper, that something remains to be done in applying the existing data. So far as the author is aware, this has been previously attempted by three writers only—Prof. R. Schöttler, Dr. A. Witz, and himself.

Before proceeding with the special consideration of the subject, it is advisable for the sake of completeness to state briefly the general laws. In doing so Rankine will be followed as closely as possible.

THERMODYNAMICS DEFINED.

'It is a matter of ordinary observation that heat, by expanding bodies, is a source of mechanical energy, and conversely, that mechanical energy, being expended either in compressing bodies or in friction, is a source of heat.

'The reduction of the laws according to which such phenomena take place to a physical theory or connected system of principles constitutes what is called the science of thermodynamics.'

FIRST LAW OF THERMODYNAMICS.

Heat and mechanical energy are mutually convertible, and heat requires for its production, and produces by its disappearance, mechanical energy in the proportion of 1,390 footpounds for each centigrade heat unit, a heat unit being the amount of heat necessary to heat one pound weight of water through 1° C. This is Joule's law, having been first determined by him in 1843. It holds with equal truth for other forms of energy, and is a general statement of the great truth, that in the universe, energy is as incapable of creation or destruction as matter. Energy may change its form indefinitely while passing from a higher to a lower level, but it can neither be created nor destroyed. The energy of outward and visible movement of matter may be arrested and caused to disappear as movement of the whole mass in one direction, but its equivalent reappears as internal movement or agitation of the particles or molecules composing the body. Energy assumes many forms, but the sum of all remains a constant quantity, incapable of change of quantity, but capable of disappearing in one form and reappearing in another.

SECOND LAW OF THERMODYNAMICS.

Although heat and work are mutually convertible and in definite and invariable proportions, yet no conceivable heat engine is able to convert all the heat given to it into work.

Apart altogether from practical limitations, a certain portion of the heat must be passed from the hot body to the cold body in order that the remainder may assume the form of mechanical energy. To get a continuous supply of mechanical energy from heat depends upon getting a continuous supply of hot and cold substances: it is by the alternate expansion and contraction of some substance, usually steam or air, that heat is converted into mechanical energy.

Perfect heat engines are ideal conceptions of machines which are practically impossible, but whose operations are so arranged that, if possible, they would convert the greatest conceivable proportion of the heat given to them into mechanical work.

Efficiency.—The efficiency of a heat engine is the ratio of the heat converted into mechanical work to the total amount of heat which enters the engine.

In this work the word *Efficiency*, when used without qualification, bears this meaning only.

The efficiency of a perfect heat engine depends upon two

things alone : these are, the temperature of the source of heat and the temperature of the source of cold (allowing the expression). The greater the difference between these temperatures the greater the efficiency. That is, the greater will be the proportion of the total heat converted into mechanical energy, and the smaller the proportion of the total heat which necessarily passes by conduction from the hot to the cold body.

Properties of Gases.—Gases are the most suitable bodies for use in heat engines ; they are almost perfectly elastic, and they expand largely under the influence of heat.

A gas is said to be perfect when it completely obeys two laws :

1. Boyle's law.
2. Charles's law.

Boyle's Law.—Suppose unit volume of gas to be contained in a cylinder fitted with a piston which is perfectly tight at unit pressure. Suppose the temperature to be kept perfectly constant. Then, according to Boyle's law, however the volume may be changed by moving the piston, the pressure is always inversely proportional to volume, that is, if volume becomes two, pressure becomes one-half ; volume becomes three, pressure becomes one-third.

The product of pressure and volume is always constant.

Denoting pressure by p, and volume by v,

Boyle's law is, $pv =$ constant.

Charles's Law.—If a gas kept at constant volume is heated, the pressure increases. If a gas is kept behind a piston which moves without friction so that the pressure upon the gas is always constant, the heat applied will cause it to expand.

One volume of gas at 0° C., if heated through 1° C. will expand $\frac{1}{273}$, and become $1\frac{1}{273}$ volume, if the pressure is constant. If the volume is constant, then its pressure will increase by $\frac{1}{273}$, that is, its pressure will become $1\frac{1}{273}$ of the original. In the same way if cooled 1° C. below 0° C., it will contract or diminish in pressure by $\frac{1}{273}$, its volume or pressure becoming $\frac{272}{273}$ of what it is at 0° C.

For every degree of heat or cold above or below 0° C. a perfect gas expands or contracts by $\frac{1}{273}$ of its volume at 0° C.

Thermodynamics of the Gas Engine

From this it is evident that a perfect gas, if cooled to 273° C. below 0° C. will have neither volume nor pressure.

This originally gave rise to the conception of absolute zero of temperature. The absolute temperature of a body is ordinary temperature Centigrade + 273, just as the absolute pressure of any gas is its pressure above atmosphere plus atmospheric pressure. The absolute temperature of a body is its temperature above Centigrade zero + 273.

The pressure or volume of a gas is therefore directly proportional to its absolute temperature.

If $p=$ pressure for absolute temperature t, and p^1 pressure for t^1 temperature, also absolute,

then $$\frac{p}{p^1} = \frac{t}{t^1};$$

or if v be the volume at absolute temperature t and v^1 at t^1,

then $$\frac{v}{v^1} = \frac{t}{t^1}.$$

The Second Law (quantitative).—If heat be supplied to a perfect heat engine at the absolute temperature T^1, and the absolute temperature of the source of cold is T, then the efficiency of that engine is, denoting it by E,

$$E = \frac{T^1 - T}{T^1} = 1 - \frac{T}{T^1}.$$

It is unity minus the lower temperature divided by the upper temperature. The efficiency is greater or less as the fraction $\frac{T}{T^1}$ is less or greater. This fraction may be diminished either by reducing T or by increasing T^1. The lowest available temperature is not capable of great variation, being in our climate about 290° absolute. It therefore follows that efficiency could only be increased by increasing T^1.

Suppose $T = 290°$ absolute and $T^1 = 580°$ absolute.

Then $$E = 1 - \tfrac{290}{580} = 1 - \tfrac{1}{2} = 0.5.$$

Suppose $T = 290°$, and $T^1 = 1450°$, a temperature common in gas engines, then

$$E = 1 - \tfrac{290}{1450} = 1 - \tfrac{1}{5} = 0.8.$$

The efficiency increases with increase of the maximum temperature. The second law, in its quantitative form, is the statement of the efficiency of any perfect heat engine in terms of absolute temperatures of the source of heat and the source of cold.

Thermal Lines.—If a volume of air is contained in a cylinder having a piston and fitted with an indicator, the piston, if moved to and fro, will alternately compress and expand the air, and the indicator pencil will trace a line or lines upon the card, which lines register the change of pressure and volume occurring in the cylinder. If the piston is perfectly free from leakage, and it be supposed that the temperature of the air is kept quite constant, then the line so traced is called an *Isothermal line,* and the pressure at any point when multiplied by the volume is a constant according to Boyle's law,

$$pv = \text{a constant.}$$

If, however, the piston is moved in very rapidly, the air will not remain at constant temperature, but the temperature will increase because work has been done upon the air, and the heat has no

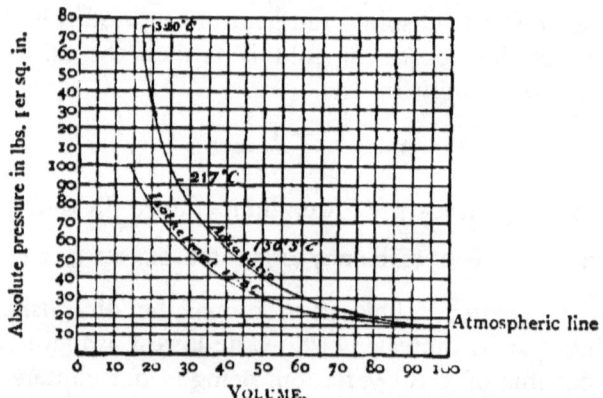

Compression lines for air (dry), Adiabatic and Isothermal.

FIG. 10.

time to escape by conduction. If no heat whatever is lost by any cause, the line will be traced over and over again by the indicator pencil, the cooling by expansion doing work precisely equalling the heating by compression. This is the line of no transmission of heat, therefore, known as *Adiabatic.* Fig. 10 shows these two lines for air starting from atmospheric pressure and temperature.

The pressures at different points of the curve are related by the equation

$$pv^y = \text{constant.}$$

The pressure when multiplied by the volume raised to the y power is always constant.

The power y is the ratio between the specific heat of the air at constant pressure and its specific heat at constant volume. According to Rankine

$$y = 1\cdot408 \text{ for air.}$$

Imperfect Heat Engines.—For a complete description of the working cycle of perfect heat engines, the reader is referred to works upon the steam engine, which contain the fullest possible details both of reasoning and results.

The working cycles of practicable heat engines are always imperfect, that is, the operations are such that, although perfectly carried out, the maximum efficiency possible by the second law of thermodynamics could not be attained by them. Each cycle has a maximum efficiency peculiar to itself, which is invariably less than $\frac{T^1 - T}{T^1}$, but which does not necessarily vary with T^1 and T.

It does not always follow that increase of the higher temperature causes increase of efficiency; conversely, it does not always follow that diminution of the upper temperature causes diminution of efficiency. Under some circumstances, indeed, the opposite effect is produced—increase of the upper temperature diminishes efficiency, while its diminution increases it, of course within certain limits.

All the gas engine cycles described in the previous chapter are imperfect in this sense, but all are practicable. It follows that if any one of them gives a higher efficiency than another in theory, it will also do so in practice, provided the practical losses do not increase with improved theory.

It is necessary before discussing the practical losses to see how the cycles compare with each other, if each be perfectly carried out. The results obtained can then be modified by examination of the way in which unavoidable practical losses affect each cycle.

Efficiency Formulæ.

If H is the quantity of heat given to an engine, and H^1 the amount of heat discharged by it after performing work, then, the portion which has disappeared in performing work is $H - H^1$, supposing no loss of heat by conduction or other cause, and the efficiency of the engine is

$$E = \frac{H - H^1}{H}.$$

Type 1.—A perfect indicator diagram of an engine of this kind is shown at fig. 11: the line abc is the atmospheric line, representing volume swept by the piston, the line ad is the line of pressures. From a to b the piston moves forward, taking in its charge, at atmospheric temperature and pressure; at b communication is instantaneously cut off, and heat instantaneously supplied, raising the temperature to the maximum, before the movement of the piston has time to change the volume. From e, the point of maximum temperature and pressure, the gases expand without loss of heat, the temperature only falling by reason of work performed till the pressure again reaches atmosphere. The curve ec is therefore adiabatic. In all cases let

- t be the initial temperature of the air in absolute degrees Centigrade.
- T the absolute temperature after explosion or heating.
- T^1 the absolute temperature of the gases after adiabatic expansion.
- p the atmospheric pressure.
- P_0 the absolute pressure of the explosion.
- v_0 the volume at atmospheric temperature and pressure.
- v the volume at the termination of adiabatic expansion.

In the particular case of diagram fig. 11, where the expansion is continued to the atmospheric line, the formula expressing the efficiency is very simple. Calling κ_v the specific heat of air at constant volume, and κ_p the sp. heat at constant pressure, then the heat supplied to the engine is

$$H = \kappa_v (T - t),$$

Thermodynamics of the Gas Engine

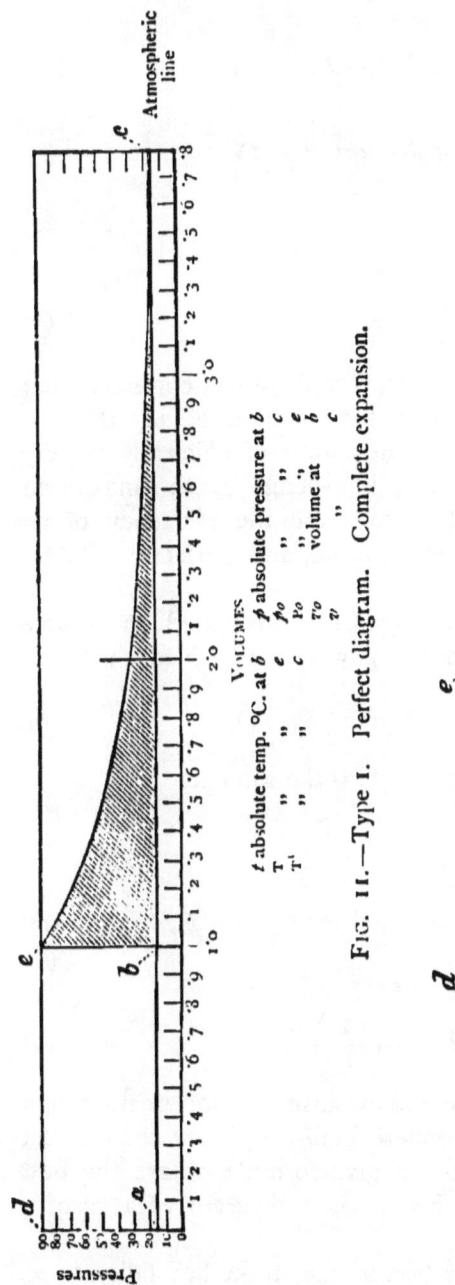

t absolute temp. °C. at b p absolute pressure at b
T ,,   ,,   ,, c p_0 ,,   ,, c
T^1 ,,   ,,   ,, e v_0 volume at b
                v ,,   ,, c

FIG. 11.—Type I. Perfect diagram. Complete expansion.

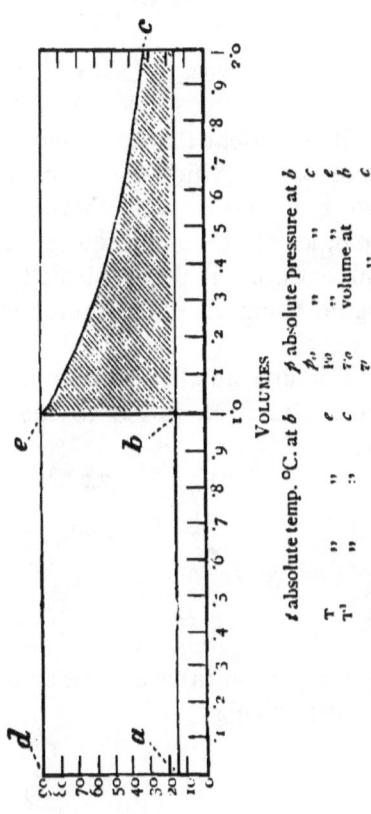

t absolute temp. °C. at b p absolute pressure at b
T ,,   ,,   ,, c p_0 ,,   ,, c
T^1 ,,   ,,   ,, e v_0 volume at b
                v ,,   ,, c

FIG. 12.—Type I. Perfect diagram. Incomplete expansion.

and the heat discharged from it is
$$H^1 = K_p(T^1 - t);$$
therefore efficiency is
$$E = \frac{K_v(T-t) - K_p(T^1-t)}{K_v(T-t)},$$
and
$$\frac{K_p}{K_v} = y$$
therefore
$$E = 1 - y\left(\frac{T^1-t}{T-t}\right). \tag{1}$$

It is evident that for every value of T there is a corresponding value of T^1, which increases with the increase of T. If T^1 is known in terms of T, then the calculation of efficiency is very rapid, as all that is required is a knowledge of the maximum temperature of the explosion to calculate the efficiency of an engine using that maximum temperature, and perfectly fulfilling this cycle.

For any adiabatic curve, the pressure multiplied by volume which has been raised to the yth power is a constant; therefore
$$P_o v_o^y = p_o v^y \text{ (see diagram, fig. 11)}, \tag{a}$$
and
$$\frac{T}{t} = \frac{P_o}{p} \text{ which, as } p = p_o, \text{ is the same as } \frac{P_o}{p_o};$$
also
$$\frac{v}{v_o} = \frac{T^1}{t}.$$

∴ in equation (a) T may be substituted for P_o, t for p_o, t for v_o, and T^1 for v, giving
$$T t^y = t T^{1y}$$
$$T^1 = t\left(\frac{T}{t}\right)^{\frac{1}{y}} \tag{2}$$

In most engines of this type the expansion is not great enough to reduce the pressure to atmosphere before opening the exhaust valve; it is therefore necessary to give formulæ where the best condition is not carried out. Fig. 12 is a diagram of a case of this kind.

The pressure at the termination of the stroke has fallen to p_o, and the temperature to T^1. The heat supplied to the engine is the same as in the first case

$$H = K_v (T - t).$$

The heat discharged by it cannot be so simply expressed. Suppose the hot gases at the pressure p_o to be allowed to cool by contact with the sides of the cylinder at constant volume till the atmospheric pressure p is reached, then the temperature

$$t^1 = T^1 \frac{p}{p_o},$$

or in terms of volume and t

$$t^1 = \frac{v}{v_o} t,$$

and the heat lost is

$$K_v (T^1 - t^1).$$

The heat to be still abstracted before the air returns to its original condition at t, and pressure p is

$$K_p (t^1 - t).$$

Total heat discharged by exhaust, therefore,

$$H^1 = K_v (T^1 - t^1) + K_p (t^1 - t).$$

The efficiency consequently is

$$\begin{aligned} E &= \frac{K_v (T - t) - \{K_v (T^1 - t^1) + K_p (t^1 - t)\}}{K_v (T - t)} \\ &= 1 - \frac{(T^1 - t^1) + y(t^1 - t)}{T - t} \end{aligned} \quad (3)$$

In this case there is no fixed relationship between T the temperature of the explosion, and T^1 the temperature of the gases at the termination of adiabatic expansion. As the expansion is more or less complete, so does T and T^1 change. In no case, however, can the efficiency be so great as that in the first case.

Type 2.—A perfect indicated diagram of an engine of this type is shown at fig. 13. Although the cycle requires two cylinders, producing two diagrams, they are better compared when superposed. The whole diagram may be supposed to come from the motor cylinder, the shaded portion of it representing the available work of the cycle, and the unshaded part, the part done by the compressing pump. The atmospheric line is *abc*. The pump volume is *ab*, the motor volume is *ac*. The pump takes in the volume *ab* at atmospheric pressure; it compresses it into an intermediate receiver, the compression line (adiabatic) is bf, passing into receiver, line fe. From the receiver it enters the cylinder at the constant pressure of compression on the line efg, supply of heat cut off at g. Then expansion (adiabatic) to the point c atmospheric pressure. The part $bfgc$ is the part available for work, the part $bfea$ representing the work of the compressing pump, which is deducted from the total motor cylinder, diagram $aegc$.

The total volume of air passed through the pump is v_o, volume swept by motor cylinder, v. So far as the heat operations are concerned, the part of the diagram to volume v'_c may be disregarded; it represents the pressing of the compressed charge into the reservoir after reaching the maximum pressure of compression (it is called v'_c because it is volume of compression). The admission to the motor cylinder is identical, so that work done in pump in that part equals work done upon the motor piston.

In addition to the letters used in type 1,

v'_c is volume of compression.
v'_p volume at point g on diagram.
p_c is pressure of compression.
t_c is temperature of compression.

The temperature, volume, and pressure letters are figured below the diagram to make matters clear. Compression is carried on from volume v_o at atmospheric pressure and temperature to volume v'_c at pressure p_c and temperature t_c, the curve being adiabatic.

After compression, heat is added without allowing the pressure to increase, but the piston moves out till the maximum temperature T is attained, and the supply of heat being completely cut off, adiabatic expansion follows till the atmospheric pressure is reached; the exhaust valve is then opened, and the hot gases discharged.

It is evident that as the pressure is constant, while heat is being given, the amount of heat given to the engine in all is

$$H = K_p (T - t_c),$$

and the heat discharged from it is also at constant pressure,

$$H^1 = K_p (T^1 - t).$$

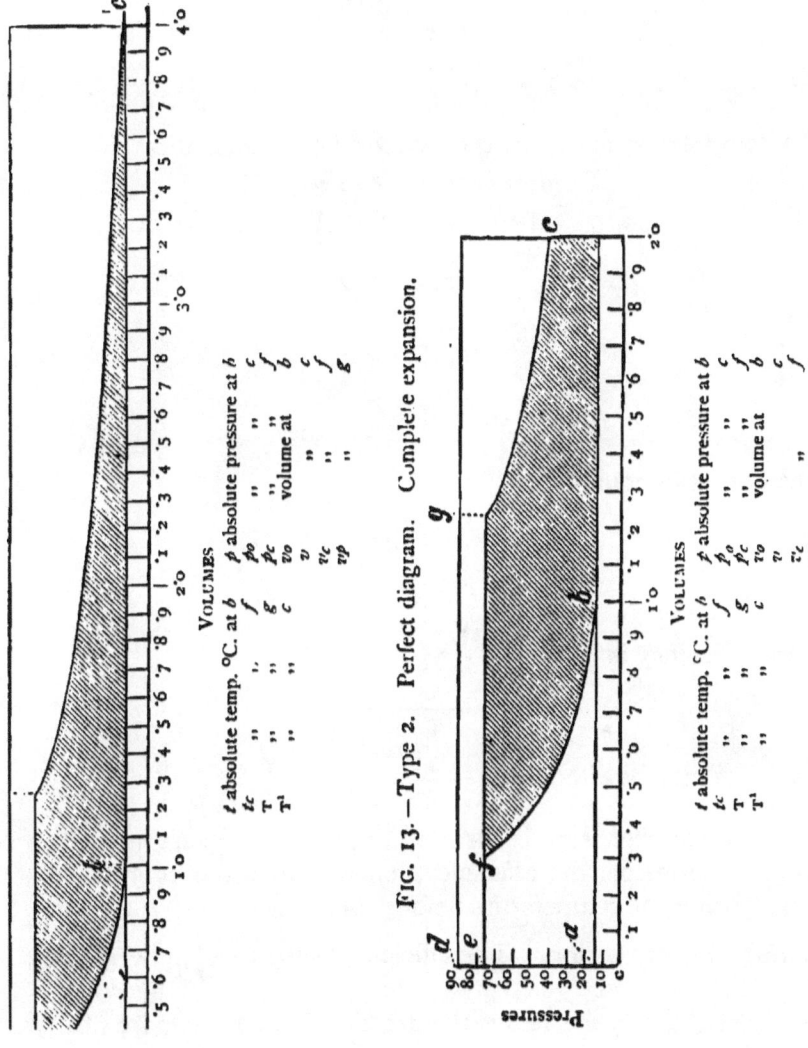

FIG. 13.—Type 2. Perfect diagram. Complete expansion.

FIG. 14.—Type 2. Perfect diagram. Incomplete expansion.

The efficiency is therefore

$$E = \frac{K_p(T - t_c) - K_p(T^1 - t)}{K_p(T - t_c)}$$

$$= 1 - \frac{T^1 - t}{T - t_c} \qquad (4)$$

The compression and expansion curves being adiabatic,

Compression $p_c v_c^y = p v_o^y$,

Expansion $p_c v_p^y = p_o v^y$;

$$\therefore \frac{v_c^y}{v_p^y} = \frac{p v_o^y}{p_o v^y}, \text{ but } p_o = p,$$

so that $\dfrac{v_c^y}{v_p^y} = \dfrac{v_o^y}{v^y}$ \qquad (a)

and $\dfrac{v_c}{v_p} = \dfrac{t_c}{T}$, also $\dfrac{v_o}{v} = \dfrac{t}{T^1}$.

Substituting in equation (a)

$$\frac{t_c}{T} = \frac{t}{T^1},$$

and $\dfrac{T^1}{T} = \dfrac{t}{t_c}$.

As the efficiency is

$$E = 1 - \frac{T^1 - t}{T - t_c},$$

it may be either $= 1 - \dfrac{T^1}{T}$ or $= 1 - \dfrac{t}{t_c}$ \qquad (5)

That is, when expansion is carried to the same pressure as existed before compression, the efficiency depends upon the compression alone, t being the temperature before compression, and t_c the temperature of compression. The efficiency being $1 - \dfrac{t}{t_c}$, the greater the temperature t_c the less is the fraction $\dfrac{t}{t_c}$, and the more nearly does E approach unity.

In most working engines of this kind, the expansion is not continued long enough to make the pressure after expanding fall to atmosphere; so that the efficiency is never so great, as when that is done, a greater portion of the heat is discharged than need be.

The modification of the formulæ is precisely as in type 1 for similar circumstances. A diagram of the kind is shown at fig. 14. The temperature t^1 is found as before:

$$t^1 = T^1 \frac{p}{p_o}.$$

The heat supplied to the cycle is as before:

$$H = K_p (T - t_c),$$

and the heat discharged is

$$H^1 = K_v (T^1 - t^1) + K_p (t^1 - t).$$

The efficiency is

$$E = 1 - \frac{\frac{1}{y}(T^1 - t^1) + (t^1 - t)}{T - t_c}. \tag{6}$$

Although there is no fixed proportion between the efficiency and the temperature of adiabatic compression, it is evident that E increases with increase of t_c.

Type 3.—A perfect indicator diagram of an engine of this type is shown at fig. 15. As in type 2, the diagrams of pump and motor are combined, the whole diagram being that given in the motor cylinder, but the shaded portion only represents the available work. The atmospheric line is *a b c*. The pump volume is *a b*, the motor cylinder volume is *a c*. The pump takes in the volume *a b* at atmospheric pressure, compresses it on the adiabatic line *b f* and into a receiver on the line *f g*. The compressed gases enter the motor cylinder on the line *g f*, heat is added instantaneously, and the pressure rises on the line *f e*. Supply of heat cut off at *e* and the expansion line *e c* is adiabatic. The total diagram in the motor cylinder is *a g f e c*, but the portion *a g f b* is common to motor and pump; the available work is therefore *b f e c*.

The total volume of air passed through the pump is v_o; the volume after adiabatic compression, from atmospheric pressure p and temperature t to pressure of compression p_c and temperature t_c is v_c. Heat is supplied at constant volume v_c till the maximum temperature of the explosion T is attained. The piston then expands the hot gases adiabatically from temperature T to T^1 and pressure P_o to pressure p_o, which in this case is equal to atmosphere.

The heat is discharged in passing from volume v to v_o at constant pressure of atmosphere. The part of the diagram from volume v_c to zero may be disregarded as it is common to both pump and motor.

The heat supplied to the cycle is
$$H = K_v (T - t_c).$$

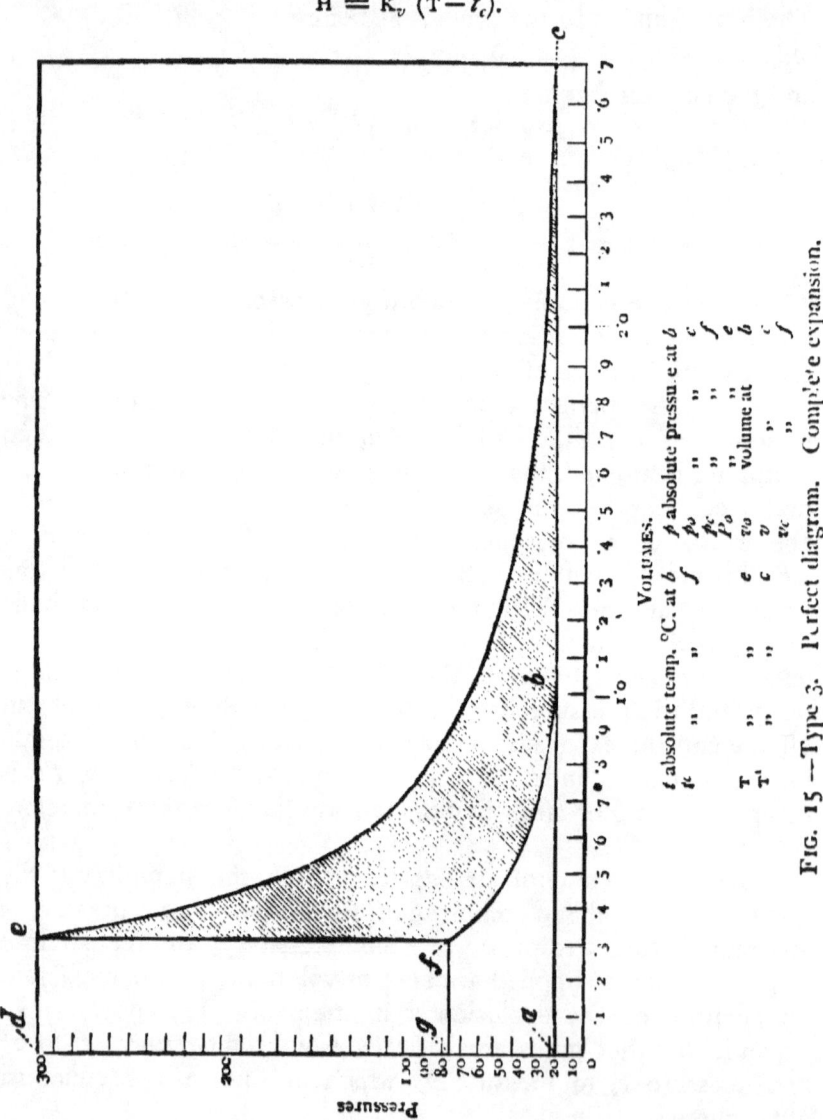

Fig. 15.—Type 3. Perfect diagram. Complete expansion.

Thermodynamics of the Gas Engine

Heat discharged
$$H^1 = K\,(T^1 - t).$$
The efficiency is
$$E = \frac{K_v(T - t_c) - K_p(T^1 - t)}{K_v(T - t_c)}$$
$$= 1 - y\frac{T^1 - t}{T - t_c}. \tag{7}$$

It is evident that for any maximum temperature T and compression temperature t_c there is a temperature T^1 at which the expansion adiabatic line falls to atmosphere. It will much simplify subsequent calculations to establish the relations between T, t_c, t and T^1.

$p_c v_c^y = p_o v^y$ and $p_c v_c^y = p v_o^y$ and as $p_o = p$,
$$\frac{p_o}{p_c} = \frac{v^y}{v_o^y}$$

but
$$\frac{v}{v_o} = \frac{T^1}{t} \text{ so that } \frac{p_c}{p_c} = \frac{T^{1\,y}}{t^y}$$

and
$$\frac{p_c}{p_c} = \frac{T}{t_c} \text{ so that } \frac{T}{t_c} = \left(\frac{T^1}{t}\right)^y.$$

T^1 in terms of T, t_c and t is therefore
$$T^1 = t\left(\frac{T}{t_c}\right)^{\frac{1}{y}}. \tag{8}$$

Although this is the best case for the third type it is not the one commonly occurring in practice; no engine has as yet been arranged to expand the gases after explosion to the atmospheric pressure.

Fig. 16 is a perfect diagram of the most common case, namely, when the expansion is carried only so far that the heat is discharged when the volume is the same as that existing before compression. The formula of efficiency is exceedingly simple, and leads to a very apparent and nevertheless somewhat paradoxical result.

The heat supplied to the cycle is
$$H = K_v\,(T - t_c),$$
and the heat discharged is
$$H^1 = K_v\,(T^1 - t),$$

because the volume of the air is the same as that existing before compression, and therefore the heat necessary to bring the fluid back to its original state can be abstracted at constant volume.

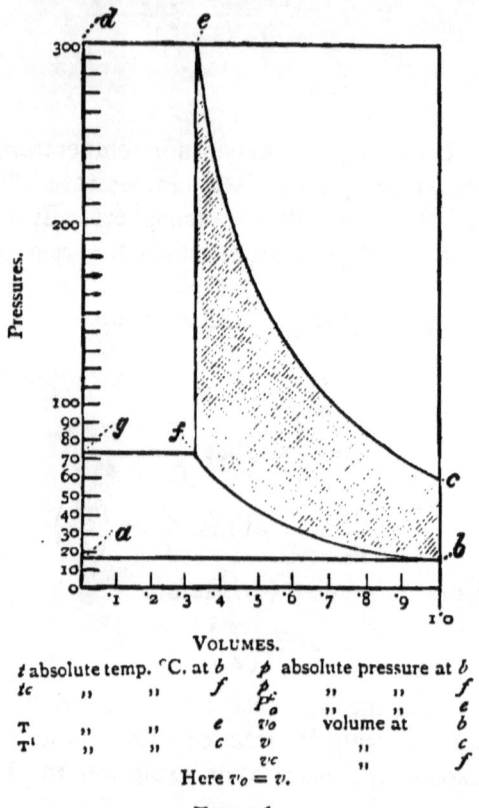

FIG. 16.

Type 3. Perfect diagram. Expansion to same vol. as before compression.

The efficiency is

$$E = \frac{K_v(T - t_c) - K_v(T^1 - t)}{K_v(T - t_c)}.$$

$$= 1 - \frac{T^1 - t}{T - t_c} \qquad (9)$$

As both curves are adiabatic, and pass through the same volume change,

$$\frac{T^1}{T} = \frac{t}{t_c};$$

so that
$$\frac{T^1 - t}{T - t_c} = \frac{T^1}{T} = \frac{t}{t_c}.$$

The efficiency may therefore be expressed

$$E = 1 - \frac{T^1}{T} \text{ or } 1 - \frac{t}{t_c} \tag{10}$$

or
$$1 - \left(\frac{v_c}{v_o}\right)^{y-1}.$$

That is, the efficiency depends upon the ratio between the initial temperature and the temperature of adiabatic compression only. T, the temperature of explosion, may be any value greater than t_a without either increasing or diminishing the efficiency. In this case

$$T^1 = \frac{Tt}{t_c}.$$

There is still another case of this type of cycle to be considered, when the expansion is continued beyond the original volume before compression, but not carried far enough to reach atmospheric pressure. Fig. 17 is a diagram of the kind.

The heat supplied to the cycle is still

$$H = K_v (T - t_c).$$

The heat discharged may be found as in a similar case with types 1 and 2.

Total heat discharged is

$$H^1 = K_v (T^1 - t^1) + K_p (t^1 - t).$$

The efficiency is

$$E = \frac{K_v (T - t_c) - \{K_v (T^1 - t^1) + K_p (t^1 - t)\}}{K_v (T - t_c)}$$

$$= 1 - \frac{(T^1 - t^1) + y (t^1 - t)}{T - t_c}. \tag{11}$$

Here then is no constant relationship between T^1 and T; the value of the cycle lies between cases 1st and 2nd. The efficiency is less than in the first case, but greater than in the second.

Type 1 A.—In this type of engine the efficiency cannot be stated in terms of temperature directly because of the nature of the perfect cycle.

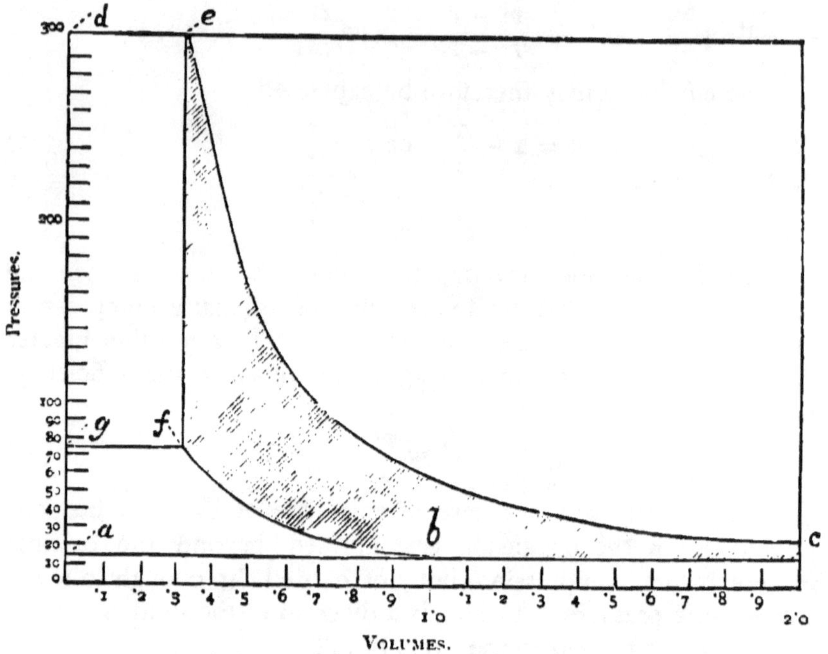

54   The Gas Engine

FIG. 17.—Type 3. Perfect diagram. Incomplete expansion.

t absolute temp. °C. at b p absolute pressure at b
t^c ,,   ,, f p_c ,,   ,, f
                    p_o ,,   ,, e
T ,,   ,, e v_o volume at b
T^1 ,,   ,, c v ,, c
                    v_c ,, f

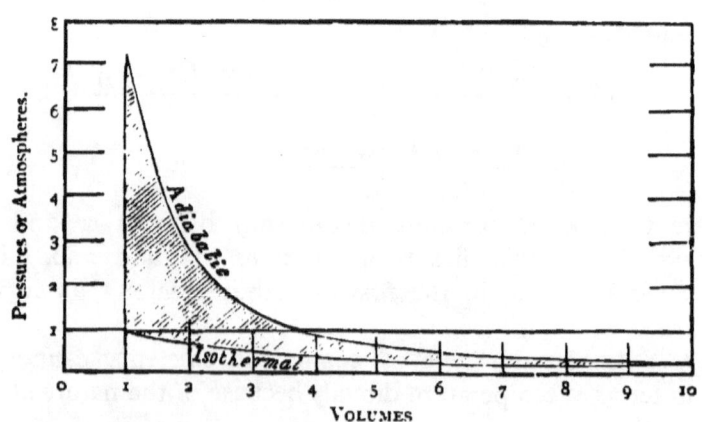

FIG. 18.—Type IA. Perfect diagram. Limited expansion.

The expansion line is adiabatic, and the compression line whereby all the heat is discharged is isothermal.[1]

Fig. 18 is the theoretical diagram of such an engine. The scale is altered from previous diagrams because of the great expansion.

There is no compression previous to the addition of heat, the heat is added at constant volume v_o, which is the volume of the charge. The pressure rises with the temperature from atmospheric pressure p and temperature t to maximum pressure P_o and temperature T. From T the expansion line is adiabatic, and is continued far enough to reduce the temperature again to t. The piston then returns, compressing the gases at the temperature t till the original volume v_o and pressure p are attained.

For any two temperatures t and T there is evidently a constant relationship between the available work and work discharged as heat. As in expanding from highest to lowest temperature the temperature falls from T to t, the whole area of the diagram $T v_o v t$, may be taken as the heat supplied to the cycle.

The heat rejected is discharged at constant temperature t, and is equivalent to the area $v_o v t t$.

For any adiabatic curve the area $T v_o v t$ is

$$\text{area} = \frac{1}{y-1}(P_o v_o - p_o v). \qquad (12)$$

For any isothermal

$$\text{area } v_o v t t = p v_o \text{ Log. } \epsilon \frac{v}{v_o}. \qquad (13)$$

The efficiency is therefore—

$$E = \frac{\frac{1}{y-1}\left(P_o v_o - p_o v\right) - \left(p v_o \text{ Log. } \epsilon \frac{v}{v_o}\right)}{\frac{1}{y-1}(P_o v_o - p_o v)}$$

$$= 1 - \frac{(y-1)\left(p v_o \text{ Log. } \epsilon \frac{v}{v_o}\right)}{P_o v_o - p_o v} \qquad (14)$$

[1] In Dr. A. Witz's able work, *Études sur les moteurs à gaz tonnant*, he falls into the error of supposing both expanding and compression lines of this type adiabatic, and he accordingly greatly over-estimates the efficiency proper to it.

but, as the line of compression discharging heat is an isothermal, that is, the temperature is kept constant at t during compression from the lowest pressure to atmosphere,

$$pv_o = p_o v \text{ (Boyle's law)}.$$

The efficiency may therefore be written

$$E = 1 - \frac{(y-1)\left(pv_o \text{ Log. } \epsilon \frac{v}{v_o}\right)}{P_o v_o - pv_o}$$

$$= 1 - \frac{(y-1) p \text{ Log. } \epsilon \frac{v}{v_o}}{P_o - p};$$

then
$$\frac{T}{t} = \frac{P_o}{p} = \left(\frac{v}{v_o}\right)^{y-1} \therefore \frac{v}{v_o} = \left(\frac{T}{t}\right)^{\frac{1}{y-1}}.$$

The efficiency can therefore be given entirely in terms of T and t:

$$E = 1 - \frac{(y-1) t \text{ Log. } \epsilon \left(\frac{T}{t}\right)^{\frac{1}{y-1}}}{T - t}. \tag{15}$$

In the case where the expansion is not carried far enough to bring the temperature of explosion down to the temperature of the atmosphere, the efficiency can be found by using the formulæ 12 and 13 to get proportions of available and total work, and then get from the nature of the compression curve the total heat discharged. As this is variable it will be better to study it from a numerical example later on.

The diagram given is the best possible for this kind of cycle.

Efficiency Formulæ for the Different Types.

The general formulæ for efficiency of the four kinds of cycle are as follows.

Type I, *1st Case*:

$$E = 1 - y \frac{T^1 - t}{T - t}. \tag{16}$$

T^1 in terms of T and t:

$$T^1 = t \left(\frac{T}{t}\right)^{\frac{1}{y}}.$$

2nd Case:
$$E = 1 - \frac{(T^1 - t^1) + y(t^1 - t)}{T - t}. \qquad (17)$$

TYPE 2, 1st Case:
$$E = 1 - \frac{T^1 - t}{T - t_c}; \qquad (18)$$

also $E = 1 - \dfrac{t}{t_c}$.

2nd Case:
$$E = 1 - \frac{\frac{1}{y}(T^1 - t^1) + (t^1 - t)}{T - t_c}. \qquad (19)$$

TYPE 3, 1st Case:
$$E = 1 - y\frac{T^1 - t}{T - t_c} \qquad (20)$$

T^1 in terms of T and t:
$$T^1 = t\left(\frac{T}{t_c}\right)^{\frac{1}{y}}.$$

2nd Case:
$$E = 1 - \frac{T^1 - t}{T - t_c}. \qquad (21)$$

also $E = 1 - \dfrac{t}{t_c}$.

3rd Case:
$$E = 1 - \frac{(T^1 - t^1) + y(t^1 - t)}{T - t_c}. \qquad (22)$$

TYPE 1 A:
$$E = 1 - \frac{t(y-1)\,\text{Log.}\,\epsilon\left(\frac{T}{t}\right)^{\frac{1}{y-1}}}{T - t}. \qquad (23)$$

Those formulæ will be found very convenient in rapidly calculating the theoretical efficiency for any kind of diagram, but they do not throw much light upon the relative advantage of the different types. In type 1, for instance, it is apparent that efficiency increases with increase of temperature because the fraction $\dfrac{T^1 - t}{T - t}$

becomes less with increase of T, but it does not rapidly become less because T^1 also increases with increase of T.

In type 2, 1st case, the efficiency is quite independent of T, and is dependent only on the ratio between t and t_c or v_o and v'_c. Increase of T (maximum temperature) increases the available portion of the engine diagram, and therefore the average pressure, but without altering the efficiency.

TYPE 3.—With this type it is easy to see (1st case) that the efficiency is greater than in type 1, but only a numerical example will show the proportion.

In the second case it may be greater or less than in type 1, depending altogether on the amount of the compression.

To obtain a clear idea of the relative values of the efficiencies, it is necessary to calculate a few numerical examples.

CALCULATED EXAMPLES OF EFFICIENCY OF THE TYPES.

Numerical Examples.—Using air as the working fluid, the value of y, the ratio of specific heat at constant volume to specific heat at constant pressure is 1·408.

$$\frac{K_p}{K_v} = y = 1 \cdot 408.$$

The gaseous mixture used in a gas engine differs considerably from pure air in its composition, and consequently in the ratio between specific heat at constant volume, and specific heat at constant pressure, but it is advisable in the first place to consider the cycle as using air pure and simple. So many circumstances modify the theoretic efficiency in actual practice that they can be best considered after studying the simpler cases.

The temperature 1600° C. is a very usual one in the cylinder of a gas engine, and it will be calculated in each instance as the maximum, 17° C. being taken as atmospheric temperature.

A similar set with 1000° C. as the maximum will be calculated to show in each case the change of efficiency, if any, with change of maximum temperature.

TYPE 1.—*1st Case.* The expansion is continued to atmospheric pressure.

Taking T = 1600° C. = 1873° absolute.
t = 17° C. = 290° „

Thermodynamics of the Gas Engine

Then T^1 = the temperature after adiabatic expansion to atmospheric pressure.

$$T^1 = t\left(\frac{T}{t}\right)^{\frac{1}{y}} \tag{2}$$

$$T^1 = 290 \left(\frac{1873}{290}\right)^{\frac{1}{1\cdot 408}} = 1090° \text{ absolute.}$$

The efficiency is

$$E = 1 - y\frac{T^1 - t}{T - t} = 1 - 1\cdot 408\,\frac{1090 - 290}{1873 - 290} = 0\cdot 29$$

$E = 0\cdot 29$ with maximum temperature of 1600° C.

Taking the maximum temperature of explosion as 1000° C.

$$\begin{aligned} &\text{Absolute} \\ T =\ & 1273° = 1000° \text{ C.} \\ t =\ & 290° =\ \ 17° \text{ C.} \\ \text{then } T^1 =\ & 829°. \end{aligned}$$

$$E = 1 - 1\cdot 408\,\frac{829 - 290}{1273 - 290} = 0\cdot 23.$$

$E = 0\cdot 23$ with maximum temperature of explosion as 1000° C.

In this cycle the efficiency evidently increases with increase of the temperature of the explosion, but not in proportion to the increase of temperature; a change of maximum temperature from 1000° to 1600° C. only causing the efficiency to rise from 0·23 to 0·29. That is, at the first temperature, 23 heat units out of every 100 given to the cycle will be converted into work, while with the second much higher temperature, only 29 units of 100 will be converted into work.

The second case of this type is the one most commonly occurring in practice. The cylinder is so arranged that the charge is taken in for half-stroke, the explosion then occurs, and the piston completes its stroke, expanding the heated gases from one volume to two volumes.

In the diagram, fig. 12, suppose volume v to be equal to $2\,v_o$, and

$$\begin{aligned} T =\ & 1873° \text{ absolute.} \\ t =\ & 290° \quad \text{,,} \end{aligned}$$

To get T^1,

$$\frac{T}{T^1} = \left(\frac{v'}{v'_o}\right)^{y-1} \text{ or } \frac{T^1}{T} = \left(\frac{v_o}{v'}\right)^{y-1}$$

$$T^1 = T\left(\frac{v^p}{v}\right)^{y-1}$$

$$T^1 = 1873\left(\frac{1}{2}\right)^{0\cdot 408} = 1411°\text{ absolute.}$$

To calculate efficiency t^1 is still required; it is, in terms of volume and t,

$$t^1 = \frac{v'}{v_o}t = \frac{2}{1}290 = 580°\text{ absolute.}$$

The efficiency can now be obtained from formula (17).

$$E = 1 - \frac{(T^1 - t^1) + y(t^1 - t)}{T - t}$$

$$= 1 - \frac{(1411 - 580) + 1\cdot 408(580 - 290)}{1873 - 290}$$

$$= 1 - \frac{831 + 1\cdot 408 \times 290}{1583} = 0\cdot 22 \text{ nearly}$$

For this case $E = 0\cdot 22$, showing the effect of limiting the expansion and discharging at a pressure above atmosphere.

Taking the same ratio of expansion and the lower maximum temperature of 1000° C.

$$T = 1273°\text{ absolute.}$$
$$t = 290° \quad ,,$$

as before, $T^1 = T\left(\frac{v^p}{v}\right)^{y-1} = 1273\left(\frac{1}{2}\right)^{0\cdot 408} = 959°\text{ absolute,}$

and t^1 is still $290 \times 2 = 580°$ absolute.

Therefore $E = 0\cdot 20$.

Here the diminution of efficiency due to diminished expansion is not so great as in the first, or rather the higher, temperature,

with complete expansion 1000° C. giving 0·23,
,, limited ,, 1000° C. ,, 0·20;
with the higher temperature of 1600° C.,
with complete expansion 1600° C. giving 0·29,
,, limited ,, 1600° C. ,, 0·22.

It is evident from these results that where the amount of expansion is from one volume to two volumes, as in the Lenoir and Hugon engines, the efficiency does not substantially improve with increasing temperature.

TYPE 2.—*1st Case.* Where the expansion is carried far enough to reduce the working pressure to atmosphere, the efficiency of this kind of engine is quite independent of the temperature of combustion. This is shown by Professor Rankine [1] in his work on the steam engine. Whether the heat added after compression be great or small in amount, the proportion of it which is converted into work is stationary.

The compression most commonly used in this kind of engine is 60 lbs. per sq. in. above atmosphere, 75 lbs. per sq. in. absolute, taking the atmospheric pressure as 15 lbs. per sq. in.

The compression is, as before stated, adiabatic; no heat is lost or gained. The temperature rises simply because of work performed upon the air.

Let

Atmospheric temperature and pressure (absolute) $t, p = 290° - 15$
Compression, „ „ „ $t_c p_c = -75$

$$\frac{t_c}{t} = \left(\frac{p_c}{p}\right)^{\frac{\gamma-1}{\gamma}}$$

$$t_c = t \left(\frac{p_c}{p}\right)^{\frac{\gamma-1}{\gamma}}$$

$$t_c = 290 \left(\frac{75}{15}\right)^{0.29} = 462.5° \text{ absolute,}$$

$$E = 1 - \frac{290}{462.5} = 0.37$$

$$E = 0.37.$$

This result is much better than any obtained with the first type. It holds equally good for all combustion temperatures; with either 1000° C. or 1600° C. the efficiency would still be 0.37, so long as that degree of compression was used. With a higher compression the efficiency increases; 100 lbs. per sq. in. above atmosphere is quite a workable degree of compression. It is instructive to calculate the efficiency with this pressure:

[1] *The Steam Engine*, Prof. Rankine, p. 373, Formula (7).

$$t = 290° \text{ absolute.}$$
$$p = 15 \text{ lbs. per sq. in. absolute.}$$
$$p_c = 115 \quad ,, \quad ,, \quad ,, \quad ,,$$
$$t_c = 290 \left(\frac{115}{15}\right)^{0.29} = 524° \text{ nearly.}$$
$$\text{E} = 1 - \frac{290}{524} = 0\cdot 45.$$
$$\text{E} = 0\cdot 45.$$

This type is evidently much superior to the first type, as it is capable of greatly increased efficiency by the mere increase of compression.

In the engines in practice expansion has not been carried far enough to give the results calculated above. It has been usual to construct the engine so that the compression pump is one-half of the volume of the motor cylinder, that is, the ratio of the expansion is from one volume to two volumes at atmosphere. Taking first a compression of 60 lbs. per sq. in. above atmosphere with this proportion between the volumes at atmosphere, and the highest temperature as 1600° C., then (diagram, fig. 13)

$$\text{T} = 1873° \text{ absolute.}$$
$$t = 290° \quad ,,$$
$$t_c = 462\cdot 5° \quad ,,$$
$$t^1 = 290 \times 2 = 580.$$

Before getting T^1 it is necessary to get the volume v_p at the highest temperature. It is

$$v_p = v_c \frac{\text{T}}{t_c}$$

and $\quad v_c = v_o \left(\dfrac{p}{p_c}\right)^{\frac{1}{\gamma}} = 1 \left(\dfrac{15}{75}\right)^{\frac{1}{1\cdot 408}} = 0\cdot 318$

$$\therefore v_p = 0\cdot 318 \frac{1873}{462\cdot 5} = 1\cdot 29$$

and $\quad \text{T}^1 = \text{T} \left(\dfrac{v_p}{v}\right)^{\gamma-1} = 1873 \left(\dfrac{1\cdot 29}{2}\right)^{0\cdot 408} = 1566° \text{ absolute.}$

The efficiency can now be found by formula (19)

$$\text{E} = 1 - \frac{\frac{1}{\gamma}(\text{T}^1 - t^1) + (t^1 - t)}{\text{T} - t_c} = 1 - \frac{\frac{1}{\gamma}(1566 - 580) + (580 - 290)}{1873 - 462\cdot 5}$$

$$= 1 - \frac{0.71\,(986) + 290}{1410.5} = 1 - 0.70 = 0.30$$
$$E = 0.30.$$

Here the insufficient expansion has caused the efficiency possible from the compression to fall from 0·37 to 0·30.

Calculating in the same way for the greater compression of 100 lbs. per sq. in. above atmosphere, with expansion ratio between compression and motor cylinders of two, it is found that the result is improved.

Here $v_c = 0.235$ vol.
and $v_p = 0.841$ vol.

$$T^1 = T\left(\frac{v_p}{v}\right)^{y-1} = 1873\left(\frac{0.841}{2}\right)^{0.408} = 1318° \text{ absolute.}$$

$$T = 1873°$$
$$t = 290°$$
$$t^1 = 580°$$
$$t_c = 524°$$

The efficiency is therefore

$$E = 1 - \frac{\frac{1}{y}(T^1 - t^1) + (t^1 - t)}{T - t_c} = 1 - \frac{0.71\,(1318 - 580) + (580 - 290)}{1873 - 524}$$

$$= 1 - \frac{0.71 \times 738 + 290}{1349} = 1 - \frac{814}{1349} = 0.40$$

$$E = 0.40.$$

The greater compression has greatly increased the efficiency while leaving the proportion of the two cylinders unaltered.

Still using the same cylinders, the efficiency with compression of 60 lbs. above atmosphere and a maximum temperature of 1000° C., is

$$E = 0.28 \text{ nearly,}$$

the data being

$T^1 = 892°$ $T = 1273°$
$t^1 = 580°$ $t = 290°$
               $t_c = 462°$,

volumes

$v_o = 1$ $v = 2$
$v_p = 0.318$ $v_p = 0.87$.

Using the higher compression 100 lbs. above atmosphere with 1000° C. as highest temperature

$$E = 0.44.$$

Data : $T^1 = 763°$ $T = 1273°$
        $t^1 = 580°$ $t = 290°$
                       $t_c = 524°$

Vol. : $v_0 = 1$ $v = 2$
        $v_1 = 0.235$ $v_p = 0.57$

In this kind of engine the best result is always obtained when the expansion is carried to atmospheric pressure. The necessary proportion between the two cylinders, to accomplish this, depends on two things : the temperature of compression, and the temperature of combustion. The ratio between the cylinders should be

$$r = \frac{T}{t_c}.$$

With a temperature of compression of 462°, for instance, and a maximum of 1873° absolute $\left(\frac{1873}{462} = 4.05\right)$ the volume of the motor cylinder would require to be 4.05 times that of the pump. With the increased compression giving 524° absolute $\left(\frac{1873}{524} = 3.57\right)$ ratio of motor to pump 3.57 to 1.

With the lower maximum temperature of 1273° the ratios for the two compression values are

$$\frac{1273}{462} = 2.75 \qquad \frac{1273}{524} = 2.43 \text{ nearly.}$$

These figures explain why the efficiency varies so much with two cylinders of ratio 1 to 2 with change of maximum temperature and compression.

TYPE 3.—*1st Case.* In this case expansion is carried to atmosphere. It is evident from the formulæ that efficiency varies to some extent with maximum temperature of the explosion.

Taking first a maximum temperature of 1600° C., as in the last type calculated, with a pressure of compression 60 lbs. above atmosphere,

The data are as follows :

Temperatures $T = 1873°$ $t = 290$
               $t_c = 462.$

Thermodynamics of the Gas Engine 65

T^1 in terms of T and t, t_i is (see p. 57)

$$T^1 = t \left(\frac{T}{t_i}\right)^{\frac{1}{\gamma}} = 290 \left(\frac{1873}{462}\right)^{\frac{1}{1.408}} = 783°$$

$$T^1 = 783°.$$

The efficiency therefore

$$E = 1 - \gamma \frac{T^1 - t}{T - t_i} = 1 - 1.408 \frac{783 - 290}{1873 - 462}$$

$$E = 0.51.$$

With compression 100 lbs. above atmosphere,

$$t_i = 524°$$

and T^1 is therefore
$$T^1 = 290 \left(\frac{1873}{524}\right)^{\frac{1}{1.408}} = 545°$$

and
$$E = 1 - 1.408 \frac{545 - 290}{1873 - 524}$$

$$E = 0.73.$$

Taking, next, 1000° C. as the highest temperature, first with the lower compression, and after with the higher compression,

with 60 lbs. compression T^1 is 595° absolute
with 100   „ T^1 is 545°   „
$E = 0.47$ at 60 lbs., $E = 0.52$ at 100 lbs., with 1000° C.

In this case the efficiency varies both with the maximum temperature of the explosion and the compression temperature previous to explosion. A glance at the numbers placed together will show clearly the relationship.

Max. temps. in °C.	1600°	1600°	1000°	1000°
Pressure of compression above atmosphere	60 lbs.	100 lbs.	60 lbs.	100 lbs.
Efficiency	0.51	0.73	0.47	0.52

2nd Case. Here the expansion after explosion is not carried on far enough to reduce the pressure to atmosphere. It terminates when the volume is the same as existed before compression, that is, the volume swept by the motor piston in expanding doing work is identical with that swept by the pump piston in compressing up to maximum pressure. Pump and motor are equal in volume. To this case of type 3 belong all compression engines in which the motor piston compresses its charge into a space at the end of the cylinder.

F

In this case, as in case 1, type 2, the theoretic efficiency of the engine is quite independent of the maximum temperature of the explosion. So long as the volume after expansion is the same as that before compression, it does not matter in the least how much heat is added at constant volume of compression; whether only a few degrees rise occurs or 1000° or 2000°, it is all the same so far as the proportion of added heat converted into work is concerned. That proportion depends solely upon the amount of compression.

For 60 lbs. adiabatic compression, temperature 462° absolute, the efficiency is 0·37; for 100 lbs. above atmosphere it is 0·45. Given by the formula $E = 1 - \dfrac{t}{t_c}$. (See p. 57.)

E depends absolutely upon the temperature of the atmosphere and the temperature of compression t and t_c. If the relative volumes of space swept by piston and compression space be known, then the efficiency can be at once calculated.

3rd Case.—Here the expansion is carried further than the original volume before compression, but not far enough to reduce the pressure to atmosphere. Efficiency is always less than in the first case with corresponding temperature of explosion and compression, but greater than in the second case. It is found by the formula:

$$E = 1 - \dfrac{(T^1 - t^1) + y(t^1 - t)}{T - t_c}$$

t^1 depends on the relationship between the volumes v_o and v the volume at atmosphere and the volume of discharge after expansion. it is always:

$$t^1 = t\,\dfrac{v}{v_o}$$

T^1 is also found by the same method as in types 1 and 2. It is better to postpone calculating any particular case of this at present, as no engine doing this has yet got into public use, and it can be considered further on in discussing the effect of increased expansion in the actual engines.

Type 1 A.—The efficiency of this type of heat cycle depends to a considerable extent upon cooling during the return stroke; in its best form, cooling at the lowest temperature during isothermal compression, it cannot be carried out without introducing the very disadvantages with which the hot-air engine was saddled, namely, a dependence upon the slow convection of air for the discharge of the heat necessarily rejected from the cycle. The rapid performance of this operation is impossible, and accordingly it is hardly fair to compare this type with those preceding; they could all of them be greatly improved in theory by introducing greater expansions and cooling by convection at the lowest temperature, but all at the expense of rate of working. The efficiency of type 1 A, will be found to be high; but it is to be kept constantly in mind that the penalty of slow rate of work was fully exacted in the practical examples of the kind in public use. They are exceedingly cumbrous, and give but a trifling power in comparison with their bulk and weight. The efficiency in this type is dependent upon T and t only.

$$E = 1 - \frac{(y-1)\,t\,\text{Log.}\,\epsilon \left(\frac{T}{t}\right)^{\frac{1}{y-1}}}{T - t}.$$

Take first
$$T = 1873°$$
$$t = 290°$$

$$E = 1 - \frac{0.408\,t \times 4.57}{1583}$$

$$E = 0.66.$$

This is a very high efficiency, but it is obtained by using an enormous expansion,

$$\frac{v}{v_o} = \left(\frac{T}{t}\right)^{\frac{1}{y-1}} = 96.7 \text{ nearly.}$$

The piston must move through nearly 100 times the original volume of the charge before the temperature is reduced to the temperature existing before igniting; in passing back to unit volume the gases must be supposed to keep at t by the cooling effect of the cylinder walls.

When $\qquad T = 1000°\,C. = 1273°$ absolute,
$\qquad\qquad t = \quad 17°\,C. = 290°\qquad$,,
the efficiency is
$$E = 0.56,$$
and the expansion required is not so great, being 37.5 volumes.

The actual ratios of expansion used in practice have not approached those proportions, and will be considered while discussing the diagrams taken from engines of this type.

COMPARISON OF RESULTS.

The two maximum temperatures used, 1600° C. and 1000° C., with the lowest temperature, 17° C., give in a perfect heat-engine, efficiencies

$$1600° C. = 0{\cdot}85 \text{ nearly,}$$
$$1000° C. = 0{\cdot}77 \text{ ,,}$$

One case in type 3 comes nearer to a perfect heat-engine than any of the others. To compare easily the following table will be useful.

TABLE OF THEORETIC EFFICIENCY.

	Max. temp. °C.	Compression		Efficiency
		Temp. abs. °C.	Pressure above atmos.	
Type 1.				
Expanding to atmosphere	1600°	—	—	0·29
,, ,, ,,	1000°	—	—	0·23
Expanding to twice volume existing before ignition	1600°	—	—	0·22
	1000°	—	—	0·20
Type 2.				
Expanding to atmosphere	—	462°	60 lbs.	0·37
,, ,, ,,	—	524°	100 lbs.	0·45
Expanding to twice volume existing before compression	1600°	462°	60 lbs.	0·30
	1600°	524°	100 lbs.	0·40
	1000°	462°	60 lbs.	0·28
	1000°	524°	100 lbs.	0·44
Type 3.				
Expanding to atmosphere	1600°	462°	60 lbs.	0·51
,, ,, ,,	1600°	524°	100 lbs.	0·73
,, ,, ,,	1000°	462°	60 lbs.	0·47
,, ,, ,,	1000°	524°	100 lbs.	0·52
Expanding to the same volume as existed before compressing	—	462°	60 lbs.	0·37
		524°	100 lbs.	0·45
Expanding to greater volume than existed before compressing, but not enough to reach atmosphere	Efficiency between 1st and 2nd cases of this type depending on ratio of expansion.			
Type 1 A.				
Expanding from max. temp. to lowest temperature	1600°	—	—	0·66
	1000°	—	—	0·56

Comparing first the best results of each type, it is evident that type 1 is the least perfect as a heat-engine, giving back only 0·29 of the total heat entrusted to it as mechanical work, and rejecting the rest of the heat. Type 2 is distinctly better, giving a maximum efficiency of 0·45, or nearly half the heat converted into work.

Type 1 A, with a heat conversion of 0·66, is still better; but type 3, with 0·73 efficiency, is best of all, coming very closely indeed to what a perfect cycle is capable of giving.

It cannot be too constantly kept in mind that it by no means follows that the best theoretic efficiency will give the best result in practice. If gained at the expense of great volume or an impracticable process, it may not be worth so much as a worse cycle where small volume of cylinder and an easy process make it more easily attainable. Type 1 A is at a great disadvantage in the matter of expansion; it requires, as has been shown, expansion of 96·7 and 37·5 volumes respectively, so great that it is practically out of comparison as a workable cycle with the others. The other cycles vary in this respect also, but the variation will fall under the consideration of mechanical efficiency at a later stage. Type 1 A is so much out that it was necessary to mention it here.

In type 1, the efficiency varies with the temperature of explosion, especially where the expansion is carried to atmosphere; the difference, however, is not great, a very large increase of maximum temperature but slightly increasing the efficiency, 1000° C. giving 0·23, and 1600° C. only 0·29, of heat conversion. When the expansion is limited to twice the volume at the moment of heating, the effect of increasing temperature in increasing the efficiency is almost nil, 1000° C. giving 0·20 efficiency, and 1600° C. only 0·22 efficiency. The conclusion to be drawn from the fact is this: in engines of the Lenoir or Hugon kind, with limited expansion, the economy is not increased by using high temperatures; a weak mixture will give as good an indicated efficiency as a strong one.

With type 2, the maximum efficiency is obtained by expanding to atmospheric pressure, and in this case it is quite independent of the temperature of combustion; it does not matter whether a great or small increase of temperature occurs at the pressure of compression, the efficiency remains the same. That is, whether much heat be added or little heat, the proportion converted into work depends on one thing only, that is, the amount of compression — the greater the compression the greater the efficiency of the engine. The pressures of compression which have been calculated, are pressures which have been used for the kind of cycle in practice. The only limits to increasing compression are the practical ones of strength of engine and leakage of piston. The difference between efficiency at 60 lbs. and 100 lbs. compression above atmosphere is considerable, the first giving $E = 0\cdot37$, the second $E = 0\cdot45$.

When the expansion is limited to twice the volume existing before compression the maximum temperature then affects the efficiency, but not to such an extent as the compression.

Type 3.—This is the best type of all from the point under consideration. The efficiency in the best form of it varies both with maximum temperature and pressure of compression. At 1600° C. maximum temperature and compression 60 lbs. per square inch above atmosphere, $E = 0\cdot51$. At the same maximum temperature but the higher compression of 100 lbs. above atmosphere, it rises to the high efficiency of $0\cdot73$, which is very nearly what a perfect heat-engine using 1600° C. and atmospheric temperature could give. With maximum temperature of 1000° C. for these two compression pressures the efficiencies are $E = 0\cdot47$ and $E = 0\cdot52$. The best case of this type is not the one occurring in practice, in fact no compression engine of this kind has ever been much which expands to atmosphere. Usually expansion is only carried to the same volume as existed before compression, and there the efficiency is quite independent of the maximum temperature; it is determined by compression solely as in type 2.

For compression 60 lbs. per square inch above atmosphere it is $0\cdot37$, and for 100 lbs. per square inch above atmosphere it is $0\cdot45$, the difference between types 2 and 3 in this case being, that type 2 expands its working fluid at the pressure of compression, which remains constant, and the pressure falls to the pressure of atmosphere by the movement of the piston doing work; in type 3 the heat is added to the working fluid at constant volume, pressure increasing, then expansion doing work, till volume before compression is attained. The one acts by increase of the volume of the working fluid by heat, the other by increase of pressure of the working fluid by heat. The one engine gives large volumes, low pressures; the other small volumes, high pressures.

In type 1 A, the change of volume required is so great that its efficiency cannot be fairly compared with the others.

Conclusions.—The best cycle for great efficiency is produced by using *compression* in the manner of type 3.

In any cycle with any definite expansion, increase of compression previous to heating produces increase of the proportion of heat converted into work. In some cases of compression cycles, increase of the highest temperature does not increase the efficiency; it may even diminish it.

There are cases in types 2 and 3 when the efficiency is quite independent of the maximum temperature, depending solely on the amount of compression employed.

CHAPTER IV.

THE CAUSES OF LOSS IN GAS ENGINES.

In calculating the efficiency of the different kinds of engines, it has been assumed that the conditions peculiar to each cycle have been perfectly complied with. In actual engines this is impossible; it is therefore necessary to discover in what manner practice fails in performing the operations required by theory.

The actual engines differ from the ideal ones in several ways:

1. The working fluid loses heat to the walls enclosing it after its temperature has been raised to the highest point;
2. The working fluid often gains heat when entering the cylinder at a time when it should remain at the lowest temperature;
3. The supply of heat is never added instantaneously as is required in some types;
4. The working fluid does not behave as a perfect gas; owing to the complex phenomena of combustion, to some extent its physical state is changed during the addition of heat;
5. The admission, transfer and expulsion of the working fluid are not accomplished without some resistance, wire-drawing during admission, back-pressure during exhaust.

The first cause of loss is by far the most considerable and will be considered first.

Loss of Heat to the Cylinder and Piston.

Although this is the most considerable source of loss in all gas engines, the stock of information in existence upon the subject is quite insufficient to justify any attempt to state a general law. So far as the author is aware, no experiments have yet been made to determine the rate at which a mass of heated air, at from 1000° to 1600° C. loses heat to the comparatively cool metal surfaces which enclose it. That the rate of flow is rapid is quite evident. Otherwise, it would be impossible to raise steam with the relatively small heating surfaces generally used in boilers. Before applying the efficiency values obtained to actual practice it is necessary to know at what rate a cubic foot of air at about 1600° C. in contact with metal walls at from 17° C. to 100° C. will lose heat; also to know how that rate changes with change of temperature and density. Much is known of the laws of cooling at lower temperatures, but little positive data exist for temperatures so high as those occurring in the gas engine. A hot gas loses heat to the colder walls enclosing it mainly by circulation or convection. The conductivity of gases for heat is very slight, and unless in some way a large surface of the gas is exposed to the cooling surface, practically no heat would escape from the working fluid in the short time during which it is exposed in gas engines. Any arrangement which favours or hastens convection will therefore increase loss by increasing the extent of hot gaseous surface exposed to the walls. The smaller the surface to which a given volume of working fluid is exposed the less heat will it lose in a given time. So far as loss of heat is concerned then, the best type of engine is that which exposes a given volume of working fluid to the smallest surface in performing its cycle. Suppose that in the three types the pistons move at the same velocity, then that which requires to move through the smallest volume, the areas of the pistons being supposed equal, will take the shortest time to perform its cycle. In the first engine the piston moves through 2·7 vols., with the hot air filling the cylinder; the second, through 3·7 vols.; and the third, through 2·4 vols. (see diagrams 11, 13 and 15). As the volumes are proportional to the time taken to perform each cycle the third type has the best of it, the time of exposure of the hot working fluid being the least: the second type is worse than the first. There is still another circumstance in addition to surface exposed and time of exposure, that is, the average temperature of the hot gas which is exposed. If the average temperature is lower in one type than in another during exposure to a given surface for a certain time, then obviously less heat will be lost in the one than in the other. Comparing the average temperatures it is found, that in the first the temperature ranges from 1600° C. to 817° C.; in the second from 1600° C. to 901° C.; and in the third from 1600° C. to 510° C. The third will therefore show a lower average temperature than the others. Three conditions are requisite in the engine which is to lose the minimum of heat from its working fluid:

1. In performing its cycle it should expose a given volume of its working fluid to the least possible cooling surface;
2. It should expose it for the shortest possible time;
3. The average temperature during the time of exposure should be as low as possible—

which conditions are best fulfilled by the third type. In addition to its advantage in theoretic efficiency it possesses the further good points in practice of proportionally small cooling surfaces, short time of exposure, and rapid depression of temperature due to work done, consequently small loss of heat to the cylinder and piston.

The diagrams, figs. 11, 13 and 15, have been selected from the others belonging to each type because the pressures, temperatures, and relative volumes closely correspond with those which would be best and at the same time readily practicable.

The flow of heat really occurring in the gas engine cylinder will be discussed when the actual diagrams come under consideration; meantime, it is sufficient to have proved that the third type will in practice give results more closely approaching its theory than the others. If in each case a constant proportion of the heat supplied were lost to the cylinder and piston, the ratio of the efficiencies would remain constant, and although it would be impossible from present data to predict the actual values, yet the relative values would be known.

Gain of Heat by the Working Fluid when Entering the Engine.

In all types of gas engine it is found most economical to keep the motor cylinders as hot as possible; they are generally worked at a temperature close upon the temperature of boiling water. This is done to diminish the loss of heat from the explosion. It follows that if the working fluid is introduced at a lower temperature it becomes heated. In the first type, the charge should be admitted and remain at the lowest temperature until the moment of explosion, which is of course impossible if the cylinder is at 100° C. As the piston itself is hotter than that, it may be supposed that the charge is heated to that point.

Taking an extreme case and calculating the effect of having an absolute temperature of 390° for the lower limit, it will be found that the efficiency is diminished. In case 1, type 1, where the expansion is carried to atmosphere with a maximum temperature of 1873° absolute = 1600° C., the value becomes reduced to 0·23.

With a maximum temperature of 1273° absolute = 1000° C the efficiency is 0·16.

TYPE I.

Initial temp. of working fluid	Max. temp.	Efficiency
17° C.	1600° C.	0·29
117° C.	1600° C.	0·23
17° C.	1000° C.	0·23
117° C.	1000° C.	0·16

Here heating, while introducing the charge will always cause diminution in efficiency, the proportion of loss being greater with the lower maximum temperature. At 1600° C. the loss is nearly one-fifth, while at 1000° C. it is close upon one-fourth.

It is very difficult to say whether it is better to work with the cylinder hot or cold. The constructor finds himself in a dilemma if the cylinder is kept as cold as the surrounding air ; then the hot gases cool more rapidly. If he keeps the cylinder hot to diminish this, the efficiency falls also. Experiment alone can decide the question.

In engines of type 2 it is a usual proceeding to leave the compression cylinder entirely without water-jacketing, under the impression that heat is thereby saved; the temperature consequently rises to very nearly that of compression, and the entering charge becomes considerably heated before compression. This is especially the case if the admission area is small, and throttling occurs ; all the energy of velocity of the entering gas becomes transformed into heat. As in the previous case the charge may be considered to rise to 117° C. before compression.

Where expansion is carried to atmosphere it has been shown that the efficiency is quite independent of the maximum temperature, but is determined by one circumstance only—the amount of the compression. As

$E = 1 - \dfrac{t^1}{t_c}$ * and t is the temperature absolute before compressing

t_c „ „ „ „ after „

and as $\dfrac{t_c}{t} = \left(\dfrac{p_c}{p}\right)^{\frac{\gamma-1}{\gamma}}$, it follows that with a constant ratio between the pressures before and after compression, the ratio of temperature before and after compressing will also remain constant; that is, the efficiency is not in any way affected by heating the working fluid, provided the same degree of compression is used. Increase of temperature previous to compression causes a proportional increase of temperature after compressing without in any way disturbing the ratio between them.

This is an important, if in appearance a somewhat paradoxical fact, and it may be stated in another way:

If an engine receives all its supply of heat at one pressure, and rejects all its waste heat at another pressure, after falling from the higher to the lower pressure by expansion doing work, the efficiency is constant for all maximum temperatures of the working fluid.

The proportion of heat converted into work is not changed in any way by increasing the temperature before compressing, and if only one degree of heat be added after compressing, the same proportion of that one degree is converted into work, as would be done with any addition of heat however great.

Where the expansion is not continued enough to reduce the pressure after heating, to atmosphere, as in the cases of this type which occur in practice, this is not quite true; the compression still remains the most powerful element of efficiency, but heating before compression produces some change, just as increase of temperature after compression produces change. The change is

* See p. 57.

not great, and it is always in the direction of improvement with a limited expansion. If the lower temperature t is increased, the compression temperature t_c increases in proportion, and is accordingly nearer the maximum temperature. The volume increases less on heating, so that the effect upon efficiency is the same as if the expansion had been increased; the terminal pressure will more closely approach atmosphere, and therefore come nearer to the condition of maximum efficiency.

In engines of type 3 the compression and expansion are often performed in the same cylinder. For this purpose it is necessary to leave at the end of the cylinder a space into which the charge is to be compressed. As the piston does not enter this space, a considerable volume of exhaust gases remains to mix with the fresh cold charge. Partly from this and partly from the heating effect of the cylinder and piston, the charge becomes considerably heated before compression. The temperature of 200° C. is not unusual. Here the simplest case is that where the expansion is continued to the same volume as existed before compression. The efficiency depends solely upon the amount of the compression; for any given degree of compression it is constant, whether the addition of heat at constant volume after compression be great or small. The efficiency is $E = 1 - \dfrac{t}{t_c}$ as in type 2 (see p. 57); and the two absolute temperatures vary in the same ratio, that is, if the charge is heated before compression, the temperature after compression will be increased in the same ratio. The two temperatures will therefore bear a constant ratio to each other, whatever the initial temperature may be, provided the compression is constant. Heating the charge before compression will consequently have no disturbing effect upon the theoretical efficiency.*

Where the expansion is carried to atmosphere the case is different. The diagram (fig. 15) may be considered to be made up of two parts giving two different efficiencies, the sum of which in this case is 0·51. In expanding from the compression volume v_c to the original volume v (compression 75 lbs. per square inch)

---

* It is here necessary to distinguish between theoretical and practical efficiency. Heating before compression diminishes efficiency in practice by increasing maximum temperature, and therefore loss of heat.

the total efficiency is 0·37, and from that volume to v and atmospheric pressure, 0·14. The latter portion still obeys the same law as in a similar case of type 1; so that if the initial temperature at volume v be supposed 117° C. it will lose efficiency in a similar way. The temperature 901° C. will still exist at that point of the expanding line, so that it may be taken as similar to the case calculated on p. 75, where 1000° C. is the maximum. The loss of efficiency there is from 0·23 to 0·16 for an initial temperature of 117° C., which makes 0·14 become nearly 0·10. The total efficiency would therefore be 0·47 instead of 0·51 without previous heating.

Efficiency diminishes with increased temperature of working fluid before compressing, if the expansion is carried to atmosphere, but does not change where the expansion is limited to the initial volume.

OTHER CAUSES OF LOSS.

The third, fourth, and fifth causes of loss require for their examination a comparison of the actual diagrams, and a knowledge of the phenomena of explosion and combustion, and so cannot be discussed at this stage.

CHAPTER V.

COMBUSTION AND EXPLOSION.

IN the preceding chapters the gas engine has been considered simply as a heat engine using air as its working fluid ; it has been assumed that in the different cycles, the engineer is able to give the supply of heat either instantaneously, or slowly, at will ; and also that he can command temperatures so high as 1000° C. or 1600° C. It is now necessary to study the properties of gaseous explosive mixtures in order to understand how far these assumptions are true.

ON TRUE EXPLOSIVE MIXTURES.

When an inflammable gas is mixed with oxygen gas in certain proportions, the mixture is found to be explosive : a flame approached to even a small volume contained in a vessel open to the air will produce a sharp detonation. Variation of the proportions will cause change in the sharpness of the explosion. There is a point where the mixture is most explosive ; at that point the inflammable gas and the oxygen are present in the quantities requisite for complete combination. After explosion the vessel will contain the product or products of combustion only, no inflammable gas remaining unconsumed, or oxygen uncombined, both having quite disappeared in forming new chemical compounds.

That mixture may be called the true explosive mixture.

Definition.—When an inflammable gas is mixed with oxygen in the proportion required for the complete combination of both gases, the mixture formed is the true explosive mixture.

If the chemical formula of an inflammable gas is known, the volume of oxygen necessary for the true explosive mixture can be at once calculated. Elementary substances combine chemically with each other in certain weights known as the atomic or combining weights: chemical symbols are always taken as representing those weights of the elements indicated. In dealing with inflammable gases used in the gas engine it is convenient to remember the following symbols and weights :

Element	Symbol	Combining weight
Oxygen	O	16
Hydrogen	H	1
Nitrogen	N	14
Carbon	C	12
Sulphur	S	32

In entering or leaving any compound the elements invariably enter or leave in weights proportional to those numbers or multiples of them. Thus hydrogen and oxygen combine with each other, forming water; the formula of the compound is H_2O, meaning that 18 parts by weight contain 16 parts of O and 2 parts of H. Similarly when carbon combines with oxygen two compounds may be formed, according to the conditions, carbonic oxide or carbonic acid, formulæ CO and CO_2, the former containing in 28 parts by weight, 12 parts of carbon and 16 parts of oxygen; the latter in 44 parts by weight containing 12 parts of carbon and 32 parts of oxygen.

The formula of a compound therefore not only indicates its nature qualitatively, but it also indicates its quantitative composition. H_2O not only tells the nature of water, but it represents 18 parts by weight; CO means 28 parts by weight of carbonic oxide: CO_2 means 44 parts by weight of carbonic acid. The numbers 18, 28 and 44 are known as the molecular weights of the three compounds in question.

When dealing with gases it is more convenient to think in volumes than in weights. It is easier, for instance, to measure the proportions of explosive mixtures by volume and to say this mixture contains one cubic inch, one cubic foot or one volume of inflammable gas to so many cubic inches, feet or volumes of oxygen.

Fortunately there exists a simple relationship between the volumes of elementary gases and their combining weights, and also between the volumes of compounds and their molecular weights.

If equal volumes of the elementary gases are weighed, under similar conditions of temperature and pressure, it is found that their weights are proportional to the combining weights. Taking the weight of the hydrogen as 1, then the weights of equal volumes of nitrogen and oxygen are 14 and 16 respectively. If then it is wished to make a mixture of hydrogen and oxygen gases in the proportion of 2 parts by weight of the former to 16 parts by weight of the latter, it is only necessary to take 2 vols. H and 1 vol. O. The law may be stated in two ways, as follows :

Taking hydrogen as unity the specific gravity of the elementary gases is the same as their combining weights ; or

The combining volumes of the elementary gases are equal.

Instead of troubling to weigh out portions of the gases it is at once known that one volume of nitrogen weighs 14 parts, the same volume of hydrogen weighing one part, oxygen 16 parts, and so on through all the gaseous elements, under the same temperatures and pressures.

Knowing that water is the compound formed by the combustion of hydrogen and oxygen, and that its formula is H_2O, it is at once apparent that the true explosive mixture of these gases is 2 vols. H and 1 vol. O. By experiment it is found that the volume of the water produced is less (of course in the gaseous state) than the volume of the mixed gases before combination.

The measurement requires to be made at a temperature high enough to keep the steam formed in the gaseous state. Measure 2 vols. H and 1 vol. O into a strong glass vessel heated to 130° C.; the total is 3 vols. ; fire by the electric spark over mercury. It will be found that the steam formed when it has cooled to 130° C. after the explosion, measures 2 vols. It has been found to be true for all gaseous compounds, that however many volumes of elementary gases combine to form them the product is always two volumes. In elementary gases, one volume always contains the combining weight ; in compound gases, two volumes always contain the molecular weight. Compared with hydrogen, therefore, the specific gravity of a gaseous compound is always one-half of the molecular weight.

As before, the law may be stated in two ways:

Taking hydrogen as unity, the specific gravity of a compound gas is half its molecular weight; or

The combining volume of a compound gas is always equal to double that of an elementary gas.

These laws are known as Gay-Lussac's laws, and form part of the very basis of modern chemistry.

Using them, the true explosive mixtures by volume and the volumes of the products of the combination can be found for any gas or mixture of gases, whether elementary or compound.

The inflammable compound gases, used in the gas engine, forming some of the constituents of coal gas are:

Inflammable gas	Formula	Molecular weight	Molecular vol.
Marsh gas	CH_4	16	2
Ethylene	C_2H_4	28	2
Carbonic oxide	CO	28	2

Applying Gay-Lussac's laws, the oxygen required for true explosive mixtures and the volumes of the products of combustion are as follows for all the inflammable gases used in the gas engine:

	H_2O Steam.	CO_2 Carbonic acid.
2 vols. hydrogen (H) require 1 vol. oxygen (O) forming	2 vols.	—
2 vols. marsh gas (CH_4) require 4 vols. oxygen (O) forming	4 vols.	2 vols.
2 vols. ethylene (C_2H_4) require 6 vols. oxygen (O) forming	4 vols.	4 vols.
2 vols. carbonic oxide (CO) require 1 vol. oxygen (O) forming	—	2 vols.
2 vols. tetrylene (C_4H_8) require 12 vols. oxygen (O) forming	8 vols.	8 vols.

With hydrogen and oxygen 3 volumes before combination become 2 volumes after combination. CH_4 and O, also C_2H_4 and O, the volumes of the products of combustion, are equal to the volumes of mixture. With carbonic oxide and oxygen 3 volumes before become 2 volumes after combination.

On Inflammability.

Previous to 1817, Sir Humphry Davy made the admirable researches which led him to the invention of the safety lamp. He then made experiments upon different explosive mixtures, and found that under certain conditions they lost the capability of ignition by the electric spark. True explosive mixtures, he observed, may lose inflammability in two ways; by the addition of excess of either of the gases or of any inert gas such as nitrogen, and by rarefaction. The hydrogen explosive mixture, if reduced to one-eighteenth of ordinary atmospheric pressure, cannot be inflamed by the spark. Heated to dull redness at this pressure it will recover its inflammability and the spark will cause combination.

One volume of the mixture to which has been added nine volumes of oxygen is uninflammable, but if the density is increased or the temperature raised, it recovers its inflammability.

Eight volumes of hydrogen added, produces the same effect as the nine volumes of oxygen, but only one volume of marsh gas or half a volume of ethylene is required. The excess which destroys inflammability varies with the temperature, increasing with increase of temperature. Heating the mixture widens the range, both of dilution with excess or inert gas and reduction of pressure.

The point where inflammability ceases by diluting is very abrupt and sharply defined. The author has found that a coal gas which will inflame by the spark in a mixture of 1 gas and 14 air will not inflame with 15 of air. If the experiment be repeated on a warmer day it may inflame with 15 of air, but will not with 16 air. As the proportion is fixed for any given temperature it will be convenient to call that proportion for any mixture the 'critical proportion.' Any mixture in the critical proportion becomes inflammable by a very small increase of temperature or pressure. The exact limits of dilution temperature and pressure have yet to be discovered.

Passing from any true explosive mixture by dilution to the mixture in the critical proportion, the inflammability slowly diminishes, the explosion becoming less and less violent, till at last no report whatever is produced, and the progress of the flame (if a glass tube is used) is easily followed by the eye.

In his great work on gas analysis, Professor Bunsen confirms Davy's observations in every particular, proving loss of inflammability by dilution and reduction of pressure as well as its restoration by heating, increase of pressure and slight addition of the inflammable gas. His work, however, was not published till 1857.

On the Rate of Flame-Propagation.

The sharp explosion of a true explosive mixture is due to the very rapid rate at which a flame, initiated at one point, travels through the entire mass and thereby causes the maximum pressure to be rapidly attained. With a diluted mixture the flame travels more slowly. Dilution therefore diminishes explosiveness in two ways—by increasing the time of getting the highest pressure and also by diminishing the highest pressure which can be got. Professor Bunsen's experiments are the earliest attempts to measure the velocity of flame movement in explosive mixtures. His method is as follows:

The explosive mixture is allowed to burn from a fine orifice of known diameter, and the rate of the current of the issuing gas carefully regulated by diminishing the pressure to the point at which the flame passes back through the orifice and inflames the explosive mixture below it. This passing back of the flame occurs when the velocity with which the gaseous mixtures issue from the orifice is inappreciably less than the velocity with which the inflammation of the upper layers of burning gas is propagated to the lower and unignited layers. Knowing then the volume of mixture passing through the orifice and its diameter, the rate of flow at the moment of back ignition is known. It is identical with the rate of flame propagation through the mixture.

Bunsen made determinations for the true explosive mixtures of hydrogen and carbonic oxide.

Velocity of Flame in True Explosive Mixtures. (*Bunsen.*)

Hydrogen mixture (2 vols. H and 1 vol. O) . . 34 metres per sec.
Carbonic oxide mixture (1 vol. CO and 1 vol. O) . 1 metre per sec. nearly.

The method is a singularly simple and beautiful one and answered thoroughly for Professor Bunsen's purpose at the time he devised it. Several objections, however, may be brought against it. The mixture in issuing from the jet into the air as flame, becomes mixed to some extent with the air and so cools down; the metal plate also, pierced with the orifice, exercises a great cooling effect. If the hole were made small enough the flame could not pass back at all, however much the flow is reduced, because the heat would be conducted away so rapidly as to extinguish the flame. This had been shown by Davy in 1817; indeed it is the principle of the safety lamp. These causes probably make Bunsen's velocities too low. MM. Mallard and Le Chatelier have made velocity determinations by a method designed to obviate those sources of error.

The explosive mixture is contained in a long tube of considerable diameter, closed at one end, open to the atmosphere at the other. At each end a short rubber tube terminates in a cylindrical space closed by a flexible diaphragm. A light style is fixed upon the diaphragms. A drum revolves close to each style, both drums upon the same shaft. A tuning fork, vibrating while the experiment is being made, traces a sinuous line upon the drum and so the rate of revolution is known. The mixture is ignited at the open end, and the flame in passing the lateral opening leading to the first diaphragm ignites the mixture there, and so moves the style and marks the drum; the arrival of the flame is signalled at the other end in the same way. The drums revolving together, the distance between the two style markings measured by the vibration marks of the tuning fork gives the time taken by the flame to move between the two points. The numbers got in this way are the rates of the communication of the flame through the mixture, back into the tube, while the flame can freely expand to the air; when both ends are closed the velocity is much greater. Then, not only does the flame spread from particle to particle of the explosive mixture at the rate due to contact of the inflamed particles with the uninflamed ones, but the expansion produced by the inflammation projects the flame mechanically into the other part and so produces an ignition, which does not travel at a uniform rate, but at a continually accelerating one. In the same way, using the open tube but firing at the closed end, the expansion of the first portion adds to the apparent velocity of propagation, and projects the last portion of the mixture into the atmosphere. The true velocity of the propagation is the rate at which the flame proceeds from particle of inflamed mixture to uninflamed particle by simple contact ; the true velocity depends upon inflammability alone, the rate under other conditions depends also upon heat evolved, and therefore movement due to expansion, mechanical disturbance of the unignited by the projection of the ignited portion into its midst. These conditions may vary much; the inflammability remains constant.

Mallard and Le Chatelier's results for the true velocity of propagations are:

VELOCITY OF FLAME IN TRUE EXPLOSIVE MIXTURES.
(*Mallard and Le Chatelier.*)

	per sec.
Hydrogen mixture (2 vols. H and 1 vol. O) . . .	20 metres.
Carbonic oxide (2 vols. CO and 1 vol. O) . . .	2·2 ,,

Bunsen's rate for hydrogen mixture seems to have been too great, and for carbonic oxide mixture too little. The rate for a true and very explosive mixture such as hydrogen is liable to be inaccurately determined, as temperature variation makes a great change, and it is difficult even with Mallard and Le Chatelier's method to obtain concordant experiments. With less inflammable mixtures the difficulty disappears. As true explosive mixtures are never used in the gas engine, their properties concern the engineer only as a preliminary to the study of diluted mixtures. The most explosive mixture which can be made with air contains a large volume of nitrogen inevitably present as diluent.

The following are some of their results with diluted mixtures, which are stated to be correct within 10 per cent. error of experiment:

VELOCITY OF FLAME IN DILUTED MIXTURES. (*Mallard and Le Chatelier.*)

		per sec.
1 vol. hydrogen mixture + ½ vol. oxygen . . .	17·3 metres.
,, ,, + 1 vol. oxygen . . .	10 ,,
,, ,, + ½ vol. hydrogen . . .	18 ,,
,, ,, + 1 vol. hydrogen . . .	11·9 ,,
,, ,, + 2 vols. hydrogen . . .	8·1 ,,

These rates show that the true explosive mixture of hydrogen and oxygen when diluted with its own volume of oxygen falls from 20 metres per second to 10 metres, that is, it becomes one-half as inflammable; when its own volume of hydrogen is the diluent, the velocity only falls to 11·9 metres per second. Hydrogen therefore has less effect in diminishing inflammability than oxygen.

Remembering the fact that the atmosphere contains one-fifth of its volume of oxygen, the remaining four-fifths being nearly all nitrogen, it is easy to get the proportions for the strongest explosive mixture possible with air. Two volumes hydrogen require 1 volume oxygen, and therefore 5 volumes air. The strongest possible mixture with air is two-sevenths hydrogen, five-sevenths air. The following experiments are for hydrogen and air in different proportions :

VELOCITY OF FLAME IN DILUTED MIXTURES. (*Mallard and Le Chatelier.*)

			per sec
Mixture, 1 vol. H and 4 vols. air	.	. . .	2 metres.
,, 1 ,, H and 3 vols. air	.	. . .	2·8 ,,
,, 1 ,, H and 2½ vols. air	.	. . .	3·4 ,,
,, 1 ,, H and 1⁴⁄₇ vols. air	.	. . .	4·1 ,,
,, 1 ,, H and 1½ vols. air	.	. . .	4·4 ,,
,, 1 ,, H and 1 vol. air	.	. . .	3·8 ,,
,, 1 ,, H and ½ vol. air	.	. . .	2·3 ,,

Very strangely the velocity is greatest when there is an excess of hydrogen present. To get just enough of oxygen for complete burning 1 volume H requires 2½ volumes air, which would be naturally supposed to be the most inflammable mixture, as it gives out the greatest heat, but for some reason it is not. When the hydrogen is increased beyond that point the velocity again falls off. A determination for coal gas and air gave 1 volume gas, 5 volumes air a velocity of 1·01 metres per second, and 1 volume gas, 6 volumes air 0·285 metres per second. With coal gas also the maximum velocity is got with the gas slightly in excess.

So far, these rates of ignition or inflammation are measures of inflammability, and are the rates for constant pressure; the rates for constant volume are very different, and the problem is a more complex one. Inflaming at the closed end of the tube, they found that even very dilute mixtures gave a sharp explosion, and in the case of hydrogen true explosive mixture, the velocity became 1000 metres per second instead of 20. With hydrogen and air 300 metres per second were obtained.

MM. Berthelot and Vieille have proved that under certain conditions even greater velocities than these are possible. The conditions, however, are abnormal, and the generation of M. Berthelot's explosive wave is exceedingly undesirable in a gas engine. It is generated by inflaming a considerable portion of the mixture at once, and so causing the transmission of a shock from molecule to molecule of the uninflamed mixture: this shock causes an ignition velocity nearly as rapid as the actual mean velocity of movement of the gaseous molecules at the high temperatures of combustion. The difference between this almost instantaneous detonation and the ordinary flame propagation may be compared to similar differences in the explosion of gun cotton discovered by Sir Frederic Abel. Gun cotton lying loosely, and open to the air, will burn harmlessly if ignited by a flame; indeed, a considerable portion may be laid upon the open hand and ignited by a flame without the smallest danger. The same quantity in the same position, if fired by a percussive detonator, will occasion the most violent explosion, the nature of the shock given to the gun cotton by the detonator causing a transmission of the kind of vibration necessary to cause its almost instantaneous resolution into its component gases.

The explosive wave in gases seems to originate in like conditions. Its velocity for the true explosive mixture of hydrogen and oxygen is 2841 metres per second, and for carbonic oxide mixture, 1089 metres per second. The velocity is independent of pressure between half an atmosphere and one and a half atmosphere. It is independent, too, of the diameter of the tube used, within considerable limits, or of the material of the tube, rubber and lead tubes giving similar results. Diluting the mixtures diminishes, and heating increases it. The experiments are very interesting and important, from a physicist's standpoint, but, fortunately for the inventor dealing with gas engines, the explosive wave is not easily generated in a gas engine cylinder; if it were, it would be impossible to run the engines without shock and hammering.

The velocity which really concerns the engineer is that due to inflammability, and expansion produced by inflaming—the velocity, in fact, with which the inflammation spreads through a closed vessel. As it cannot be discussed without considering other matters—heat evolved by combustion, and temperatures and pressures produced—it will be advisable first to give the heat evolved by combustion, and then devote a complete chapter to explosion in a closed vessel.

HEAT EVOLVED BY COMBUSTION.

Careful experiments upon the heat evolved by the combustion of gases in oxygen have been made by Favre and Silberman, and also by Professor Andrews. The physicists first named burned the gases at constant pressure in a specially devised calorimeter. Professor Andrews mixed the gases in a thin spherical copper vessel, closed it, and exploded by the spark: the vessel being surrounded by water gave up its heat to the water, the weight of which being known, the rise of temperature gave the heat evolved.

Quantities of heat are measured by taking water as the unit. In this work, a heat unit always means the amount of heat necessary to raise unit weight of water through 1° C.

Taking an average of Favre and Silberman and Andrews's results, the inflammable gases used in gas engines evolve upon complete combustion the following amounts of heat:

	Heat units.
Unit weight of hydrogen completely burned to H_2O evolves	34,170
Unit weight of carbon completely burned to CO_2 evolves	8,000
Unit weight of carbonic oxide completely burned to CO_2 evolves	2,400
Unit weight of marsh gas completely burned to CO_2 and H_2O evolves	13,030
Unit weight of ethylene completely burned to CO_2 and H_2O evolves	11,900

That is, one pound weight of hydrogen burned completely to water will evolve as much heat as would raise 34,170 lbs. of water through 1° C., or the converse. One pound of carbon in burning to carbonic acid evolves as much heat as would raise 8,000 lbs. of water through 1° C. These numbers give the amount or quantity of heat evolved. The intensity or temperature of the combustion may be calculated on the assumption that the whole heat is evolved under such conditions that no heat is lost, or is applied to anything else but the products of combustion. To make the calculation it is necessary to know the specific heat of the products.

The amount of heat required to heat unit weight of water through one degree is 1 heat unit, the specific heat of any other body is the number of heat units required to heat unit weight of the body through one degree. Gases have two different specific heats depending upon whether heat is applied while the gas is kept at constant volume, or at constant pressure; both are required in dealing with gas engine problems. The specific heat at constant volume is sometimes known as the true specific heat; in taking the specific heat at constant pressure the gas necessarily expands, and so does work on the external air; this specific heat is therefore greater than the former by the amount of work done. For the gases used in the gas engine the two values are as follows. The ratio between the two is also given, as it is frequently required in efficiency calculations. The experimental numbers are Regnault's, the calculated specific heat at constant volume, Clausius.

SPECIFIC HEATS OF GASES.

(For equal weights. Water = 1.)

Name of gas	Sp. heat at constant pressure	Sp. heat at constant volume	Sp. heat con. pres. Sp. heat con. vol.
Air	0·237	0·168	1·413
Oxygen	0·217	0·155	1·403
Nitrogen	0·244	0·173	1·409
Hydrogen	3·409	2·406	1·417
Marsh gas	0·593	0·467	—
Ethylene	0·404	0·332	1·144
Carbonic oxide	0·245	0·173	1·416
Steam	0·480	0·369	1·302
Carbonic acid	0·216	0·171	1·165

It is convenient to remember that the specific heats of combining or atomic weights of the elements are equal—Dulong and Petit's law. To this law there are few exceptions, and the permanent elementary gases, oxygen, nitrogen, and hydrogen, obey it almost absolutely. As equal volumes of these gases represent the combining weights, it follows that equal volumes of these gases have the same specific heat. Taking the specific heat of air as the unit, the specific heat of hydrogen and oxygen gases is also unity. The compound gases do not obey the law so closely. The calculation of temperature of combustion can now be made. The amount of heat evolved from unit weight of a combustible is usually said to measure its calorific power, that amount divided by the specific heat of the products of the combustion is said to be the measure of its calorific intensity. The calorific intensity is indeed the theoretical temperature of the combustion: taking hydrogen first, unit weight evolves 34,170 heat units. But the water formed weighs 9 units (from formula H_2O), and if its specific heat in the gaseous state were unity, the supposed maximum temperature of combustion would be $\frac{34170}{9} = 3796·6$. But the specific heat is less than unity; therefore the theoretical maximum will be greater. It is $\dfrac{34170}{9 \times 0\cdot 480} = 7909\cdot 7$. For certain reasons to be considered later, no such enormous temperatures are ever attained by combustion. In the above calculation the latent heat of steam should first have been deducted, as it is included in the total heat evolved as measured by the calorimeter: it is 537 heat units. $34{,}170 - 537$ gives the total heat available for increasing the temperature, the amended calculation is $\dfrac{34170 - 537}{9 \times 0\cdot 480} = 7785\cdot 4$, still an exceedingly high temperature.

Calculating the heat evolved by burning carbon in the same way, but omitting any deduction for the latent heat of carbonic acid (it does not affect the calorimeter, as it does not condense), the theoretical temperature produced by burning in oxygen is still higher, being $10{,}174°$ C. Burning in air the theoretical temperatures are lower as the nitrogen present acts as a diluent, and must necessarily be heated to the same temperature as the products of the combustion. They are given as follows in 'Watts' Dictionary.'

	Calorific power	Temperature produced	
		In oxygen	In air
Carbon	8080	10174° C.	271 C.
Hydrogen	34462	6930° C.	2741° C.

These are the supposed temperatures burning in the open atmosphere, and therefore at constant pressure, the gases expanding doing work upon the air. At constant volume, that is, burning in a closed vessel so that the volume cannot increase but only the pressure, the temperature should be greater as the specific heat at constant volume is less. Allowing for that, the numbers become

THEORETICAL TEMPS. OF COMBUSTION AT CONSTANT VOLUME.

	Temperature produced	
	In oxygen	In air
Carbon	12820	—
Hydrogen	9010	4119

Such temperatures have never been produced by combustion,

for many reasons, of which all save the most potent have been discussed by the earlier writers on heat. This is Dissociation.

DISSOCIATION.

Most chemical combinations, while in the act of formation from their constituent elements, evolve heat, and as a general rule, the greater the heat evolved the more stable is the compound formed. The compound after formation may generally be decomposed by heating to a high enough temperature, heat being one of the most powerful splitting up agencies known to the chemist. The nature of the decomposition varies with the compound. In many cases the process is irreversible, that is, although heating up will cause decomposition, cooling down again, however slowly, will not cause recombination. In some compounds, however, under certain conditions the process is reversible, and recombination occurs on slow cooling.

Definition.—Dissociation may be defined as a chemical decomposition by the agency of heat, occurring under such conditions that upon lowering the temperature the constituents recombine.

Groves found long ago that water begins to split up into oxygen and hydrogen gases at temperatures low compared to that produced by combustion. Deville made a careful study of the phenomena, and found that decomposition commences at 960° to 1000° C. and proceeds to a limited extent : raising the temperature to 1200° C. increases it, but a limit is reached. The amount of decomposition depending upon the temperature, for each temperature there is a certain proportion between the amount of steam and the amount of free oxygen and hydrogen gases present. If the temperature is increased, the proportion of free gases also increases : if temperature is diminished, the proportion of free gases diminishes. If the temperature be raised beyond a certain intensity, the water is completely decomposed : if lowered beyond a certain temperature, complete combination results. The same thing happens with carbonic acid, the temperature of decomposition is lower.

It is quite evident, then, that at the highest temperatures produced by combustion, the product cannot exist in the state of complete combination. It will be mixed to a certain extent with the free constituents which cannot combine further until the temperature falls; as the temperature falls, combustion will continue till all the free gases are combined. The subject, from its nature, is a difficult one in experiment, and accordingly different observers do not quite agree upon temperatures and percentages of dissociation, but all are agreed that dissociation places a rigid barrier in the way of combustion at high temperatures, and prevents the attainment of temperatures, by combustion, which are otherwise quite possible. With no dissociation, hydrogen burning in oxygen should be able under favourable circumstances to give a temperature of over 6000° C., as has been shown. Deville's experiments upon the temperature of the oxyhydrogen flame, at constant pressure of the atmosphere, gave under 2500° C. The estimate was made by melting platinum in a lime crucible, with the oxyhydrogen flame playing upon the platinum, the crucible being well protected against loss of heat by lime blocks, so that the platinum could really attain the temperature of the flame; when at the highest temperature, the molten platinum was rapidly poured into a weighed calorimeter, and the rise in temperature noted. From this was calculated the temperature of the platinum. The experiment was dangerous and inaccurate, but it is the only serious attempt which has been made to determine the temperature of the oxyhydrogen flame at constant pressure.

The highest temperature produced by hydrogen burning in oxygen has been determined by Bunsen, and also Mallard and Le Chatelier, for combustion at constant volume, that is, explosion.

As the theoretic calculation shows, with no dissociation a temperature of 9000° C. is possible. The highest maximum it is possible to assume from Bunsen's experiments is 3800° C.; from Mallard and Le Chatelier's, 3500° C. The two sets of experiments are concordant. It is true the latter physicists do not attribute the difference wholly to dissociation, but they agree that part is due to this cause; and that there is an enormous difference between heat temperature actually got and that which should be possible if no limit existed all are agreed. With air, Bunsen's figures show a maximum of about 2000° C., Mallard and Le Chatelier say 1830° C.; the present writer has also made experiments with hydrogen in air, and finds the highest possible temperature to be 1900° C. The calculated maximum is 3800° C. The difference is not so great as with the true explosive mixture, which is to be expected, but all experiments agree in proving that there is a considerable difference.

CHAPTER VI.

EXPLOSION IN A CLOSED VESSEL.

THE value of any inflammable gas for the production of power by explosion, can be determined apart altogether from theoretical considerations by direct experiment. It is evident that the gas which for a given volume causes the greatest increase in pressure, will give the greatest power for every cubic foot used, provided that the pressure does not fall so suddenly that it is gone before it can be utilised by the piston.

Two qualities will be possessed by the best explosive mixture : (1) greatest pressure per unit volume of gas: (2) longest time of maximum pressure when exposed to cooling.

In the gas engine itself the conditions are so complex that the problem is best studied in the first instance under simplified conditions. The author has made a set of experiments upon many samples of coal gas mixed with air in varying proportions, to find the pressures produced, and the duration of those pressures; igniting mixtures at atmospheric pressures and temperature, and also at higher temperature and initial pressures. He has made some experiments upon pure hydrogen and air mixtures in the same apparatus for comparison.

The experimental apparatus is shown at fig. 19. It consists of a closed cylindrical vessel 7 inches diameter and $8\frac{1}{4}$ inches long, internal measurement, and therefore of 317 cubic inches capacity. It is truly bored, and the end covers turned so that the internal surface is similar to that of an engine cylinder ; the covers are bolted strongly so as to withstand high pressures. Upon the upper cover is placed a Richards indicator, in which the reciprocating drum has been replaced by a revolving one; the rate of revolution is adjusted by a small fan, a weight and gear giving the power.

The cylinder is filled with the explosive mixture to be tested; the drum is set revolving, the pencil of the indicator pressed gently against it, and the electric spark is passed between the points placed at the bottom of the space. The drum is enamelled and the pencil is a black-lead one. The pressure of the explo-

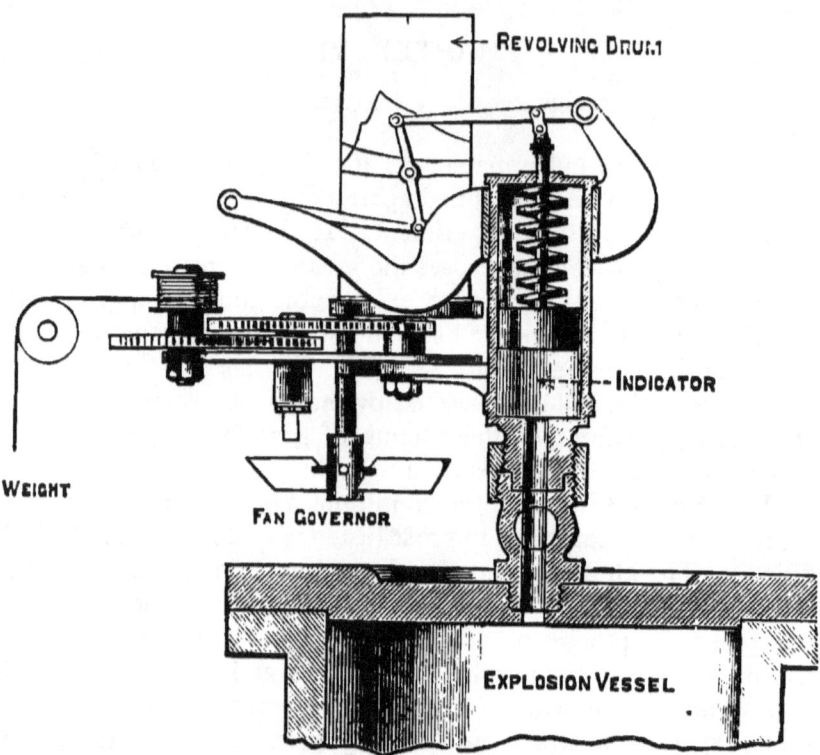

FIG. 19.—Clerk Explosion Apparatus.

tion acts upon the indicator piston, and a line is traced upon the drum, which shows the rise and fall of pressure. The rising line traces the progress of the explosion; the falling line the progress of the loss of pressure by cooling. The rate of the revolution of the drum being known, the interval of time elapsing between any two points of the explosion or cooling curve is also known. That is, the curve shows the maximum pressure attained, the time of attaining it, and the time of cooling. Line *b* on fig. 20 is a facsimile of the curve produced by the explosion of a mixture containing 1 vol. hydrogen and 4 vols. air. Each revolution of the drum was accomplished in 0·33 sec., so that each tenth of a revolution takes 0·033 sec. The vertical divisions give time; the horizontal, pressures. In this experiment the maximum pressure produced by the explosion is 68 lbs. per square inch above atmosphere, and it is attained in 0·026 second. Compared with the rate of increase the subsequent fall is very slow. The rise occurs in 0·026 second; the fall to atmosphere again takes 1·5 second, or nearly sixty times the other. It is in fact an indicator diagram from an explosion where the volume is constant, the motor piston being absent, and the only cause of loss of pressure is cooling by the enclosing walls. The exact composition of the mixture, its uniform admixture, the temperature and pressure before ignition, are all accurately known. After studying explosions under these known conditions, it becomes easier to understand what occurs under more complex conditions, where the moving piston makes the cooling surface change, and where the expansion doing work also requires consideration. As the rapidity of the increase of pressure measures the explosiveness of a mixture, the time occupied from the commencement of increase to maximum pressure will be called the *time of explosion*. The explosion is complete when maximum pressure is attained. It does not follow from

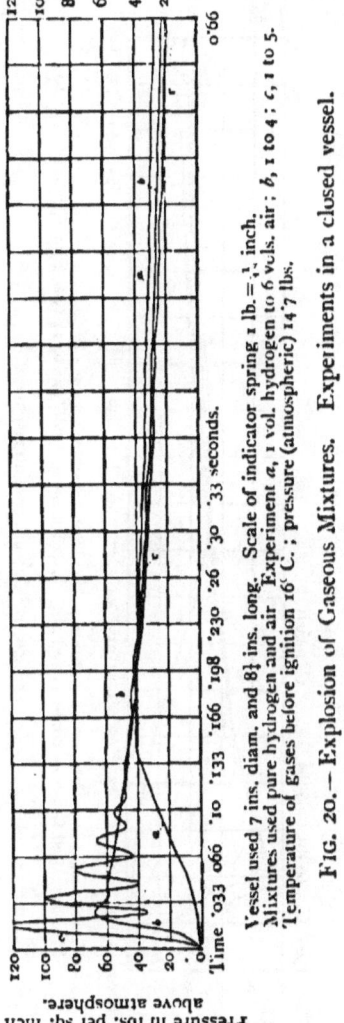

FIG. 20.—Explosion of Gaseous Mixtures. Experiments in a closed vessel.

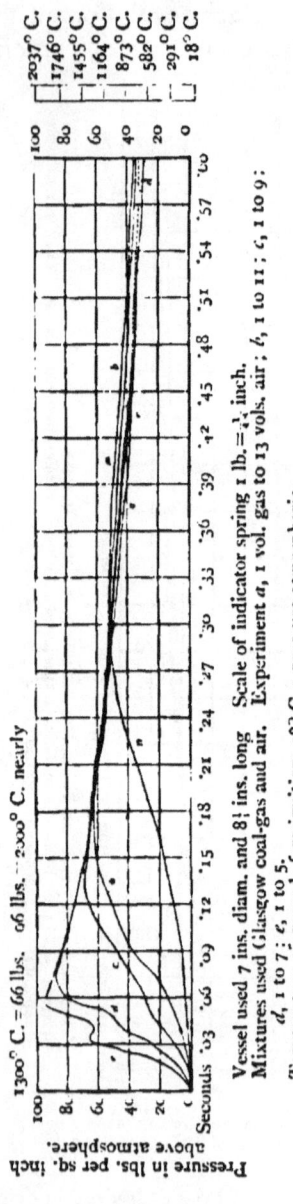
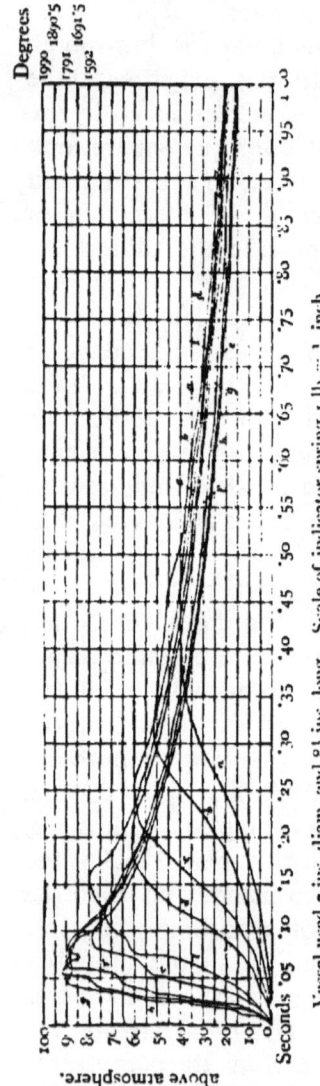

FIG. 21.—Explosion of Gaseous Mixtures. Experiments in a closed vessel.

this that the combustion is complete; that is another matter. The explosion arises from the rapid spreading of the flame throughout the whole mass of the mixture, which may be called the inflammation of the mixture.   More or less rapid inflammation means more or less explosive effect, but not complete combustion. The complete burning of the gases present does not occur till long after complete inflammation.

The terms *combustion, explosion,* and *inflammation* will be used in this sense alone :

Combustion, burning ; complete combustion, the complete burning of the carbon of the combustible gas to carbonic acid, and the hydrogen to water. So long as any portion of the combustible remains uncombined with oxygen the combustion is incomplete.

Complete explosion, the attainment of maximum pressure.

Time of explosion; the time elapsing between beginning of increase and maximum pressure.

Complete inflammation, the complete spreading of the flame throughout the mass of the mixture.

Confusion has arisen through the indifferent use of these terms, which are really distinct and are not synonymous.

With mixtures made with Glasgow coal gas the author has obtained the following maximum pressures and times of explosion.

EXPLOSION IN A CLOSED VESSEL. (*Clerk.*)
Mixtures of air and Glasgow coal gas.

Temp. before explosion . . . . . . 18° C.
Pressure before explosion . . . . . . atmospheric.

Mixture		Max. press. above atmos. in pounds per sq. in.	Time of explosion
Gas.	Air.		
1 vol.	13 vols.	52	0·28 sec.
1 vol.	11 vols.	63	0·18 sec.
1 vol.	9 vols.	69	0·13 sec.
1 vol.	7 vols.	89	0·07 sec.
1 vol.	5 vols.	96	0·05 sec.

The highest pressure which any mixture of coal gas and air is capable of producing without compression is only 96 lbs. per sq. in. above atmosphere and the most rapid increase is not more rapid than always occurs in a steam cylinder at admission. Many are still prejudiced against gas, compared with steam, because of the so-called explosive effect, and the fear that gas explosions may occasion pressures quite beyond control, like solid explosives. The fear is quite unfounded; the pressure produced by the strongest possible mixture of coal gas and air is strictly limited by the pressure before ignition, and can always be accurately known: and so provided for by a proper margin of safety in the cylinders and other parts subject to it.

The most dilute mixture of air and Glasgow gas which can be ignited at atmospheric pressure and temperature contains $\frac{1}{14}$ of its volume of gas, and the pressure produced is 52 lbs. above atmosphere. The time of explosion is 0·28 second; so slow is the rise that it cannot with justice be termed an explosion. It is too slow to be of any use in an engine running at any reasonable speed; the stroke would be almost complete before the pressure had risen. The mixture containing 6 volumes of gas is that with just enough oxygen to burn the gas. It is anomalous that the highest pressure is given with excess of coal gas. The rate of ignition also is greatest with that mixture. This agrees with the results obtained by Mallard and Le Chatelier, excess of hydrogen giving the highest rate of inflammation.

Similar experiments were made with air and Oldham coal gas.

EXPLOSION IN A CLOSED VESSEL. (*Clerk*.)
Mixtures of air and Oldham coal gas.

Temp. before explosion . . . . . . 17° C.
Pressure before explosion . . . . . atmospheric.

Mixture		Max. press. above atmos. in pounds per sq. in.	Time of explosion
Gas.	Air.		
1 vol.	14 vols.	40	0·45 sec.
1 vol.	13 vols.	51·5	0·31 sec.
1 vol.	12 vols.	60	0·24 sec.
1 vol.	11 vols.	61	0·17 sec.
1 vol.	9 vols.	78	0·08 sec.
1 vol.	7 vols.	87	0·06 sec.
1 vol.	6 vols.	90	0·04 sec.
1 vol.	5 vols.	91	0·055 sec.
1 vol.	4 vols.	80	0·16 sec.

The highest pressure in this case is 91 lbs. per square inch above atmosphere, but the most rapid explosion is 0·04 second and 90 lbs. pressure, a little less pressure than is given by Glasgow gas but a slightly more rapid ignition. The mixtures are evidently more inflammable, as the critical mixture is $\frac{1}{15}$ volume of gas instead of $\frac{1}{15}$ as with Glasgow gas. Although repeatedly tried, a mixture of 1 volume gas 15 volumes air failed to inflame with the spark.

Hydrogen and air mixtures were also tested as follows :

EXPLOSION IN A CLOSED VESSEL. (*Clerk.*)
Mixtures of air and hydrogen.

Temp. before explosion . . . . . . 16° C.
Pressure before explosion . . . . . atmospheric.

Mixture		Max. press. above atmos. in pounds per sq. in.	Time of explosion
Hyd. 1 vol.	Air. 6 vols.	41	0·15 sec.
1 vol.	4 vols.	68	0·026 sec.
2 vols.	5 vols.	80	0·01 sec.

The inferiority of hydrogen to coal gas, volume for volume, is very evident; the highest pressure is only 80 lbs. above atmosphere, and the mixture requires $\frac{2}{7}$ of its volume of hydrogen to give it, while coal gas gives the same pressure with about $\frac{1}{16}$ volume. The hydrogen mixture, too, ignites so rapidly that it would occasion shock in practice, the strongest mixture having an explosion time of one-hundredth of a second. With gas the most rapid is four-hundredths of a second.

THE BEST MIXTURE FOR USE IN NON-COMPRESSION ENGINES.

From these tables can be ascertained the best gas and the best mixture for use in non-compression engines with cylinders kept cold. Take first Glasgow gas, and determine which mixture gives the best result.

(1) Power of producing pressure.

Suppose one cubic inch of Glasgow coal gas to be used in each of the five mixtures, whose maximum pressures and times of explosion are given in the table on p. 99, the mixtures would measure respectively 14, 12, 10, 8, and 6 cubic inches. Let them be placed in cylinders of 14, 12, 10, 8 and 6 square inches piston area : the piston will in each case be raised one inch from the bottom of its cylinder. If the pressures upon the piston were the same, equal movements of piston would give equal power ; if therefore the mixtures gave equally good results the maximum pressure multiplied by the piston area will in all cases be the same.

Multiplying 14, 12, 10, 8 and 6 by their corresponding pressures 52, 63, 69, 89, and 96 respectively, the products are 728, 756, 690, 712, and 576. These numbers are the pressures in pounds which each mixture is capable of producing with one cubic inch of Glasgow coal gas, cylinders of such area being used that the depth of mixture is in every case one inch.

Proportion of Glasgow gas in mixture	$\frac{1}{14}$, $\frac{1}{12}$, $\frac{1}{10}$, $\frac{1}{8}$, $\frac{1}{6}$.
Pressure produced upon pistons by one cubic inch . . .	728, 756, 690, 712, 576 pounds.

The best mixture is seen at a glance ; it is that containing one-twelfth of gas. The pressure produced by one cubic inch of gas is at its highest value 756 pounds, in a cylinder of 12 inches piston area, and containing 12 cubic inches of mixture.

In modern gas engines the time taken by the piston to make the working part of its stroke is generally about one-fifth of a second. If the pressure in one mixture has fallen more, proportionally in that time, then although it may give the highest maximum, it may lose too rapidly to give the highest mean pressure. To find this cooling effect, find the pressure to which each mixture falls at the end of 0·2 second after maximum pressure ; it is in the different cases :

Mixture containing gas . . .	$\frac{1}{14}$, $\frac{1}{12}$, $\frac{1}{10}$, $\frac{1}{8}$, $\frac{1}{6}$.
Time after beginning explosion (0·2 sec. after max. pressure) .	0·48, 0·38, 0·33, 0·27, 0·25 sec.
Pressure in lbs. per sq. in . .	43, 48, 47, 55, 57.
Press. respectively by 14, 12, 10, 8, and 6	602, 576, 470, 440, 342.

The lower row expresses the relative pressures still remaining after allowing each explosion to cool for one-fifth of a second from complete explosion ; they express the resistance to cooling possessed by the mixtures. It is evident at once that the strongest mixtures cool most rapidly; a higher temperature being produced, more of the heat of the explosion is lost in a given time.

(2) Power of producing pressure and resisting cooling.

To find the best mixture for producing pressure and resisting cooling, those numbers are to be added to the corresponding ones for maximum pressure :

Proportion of Glasgow gas in mixture		$\frac{1}{14}$, $\frac{1}{12}$, $\frac{1}{10}$, $\frac{1}{8}$, $\frac{1}{6}$.
Pressure produced upon pistons by one cubic inch gas	. . .	728, 756, 690, 712, 576.
Pressure remaining upon pistons 0·2 sec. after complete explosion	.	602, 576, 470, 440, 342.
Mean pressure		665, 666, 580, 576, 459.

The mean of the two sets gives numbers expressing the relative values of the mixture for producing pressure, and at the same time resisting cooling. The two weakest mixtures are best in both respects, the low result given by the strongest mixture is due to the fact that excess of gas is present and it remains unburned, it proves how easily the consumption of an engine may be increased by even a slight excess of gas in the mixture.

The two best mixtures ignite too slowly, but in the actual engine that is easily controlled, as will be explained later. The best mixtures are 1 vol. gas 13 volumes air, and 1 vol. gas 11 volumes air. With more gas the economy will rapidly diminish.

The experiments with Oldham gas treated in the same way give the following results :

Proportion of Oldham gas in mixture	. . .	$\frac{1}{18}$, $\frac{1}{14}$, $\frac{1}{13}$, $\frac{1}{12}$, $\frac{1}{10}$, $\frac{1}{8}$, $\frac{1}{7}$, $\frac{1}{6}$, $\frac{1}{5}$.
Pressure produced upon pistons by one cubic inch gas.	.	600, 721, 780, 732, 780, 696, 630, 546, 400.
Pressure remaining upon pistons 0·2 sec. after complete explosion per sq. inch.	. .	31, 40, 4, 44, 44, 47, 52, 50, 46.
Pressure per piston	. . .	465, 560, 546, 528, 440, 376, 364, 300, 230.
Mean pressure upon piston	.	532, 640, 663, 630, 610, 536, 497, 423, 315.

Here, too, the best mixture lies between one-twelfth and one-fourteenth of gas ; with less and more gas the result becomes worse and worse. Glasgow and Oldham gases seem to be very nearly equal in value per cubic foot for the production of power, as the pressure produced from one cubic inch in the best mixture of each is very similar. The average pressures during 0·2 second from complete explosion are exceedingly close, Glasgow gas mixture containing one-twelfth gas giving 666 lbs. pressure per cubic inch of gas, and Oldham gas for the same mixture and the same quantity giving 630 lbs.: Glasgow gas one-fourteenth mixture 665 lbs. pressure, Oldham gas 640 lbs. The hydrogen experiments give as follows :

Proportion of hydrogen gas in mixture .	$\frac{1}{4}$,	$\frac{1}{5}$,	$\frac{2}{7}$.
Pressure produced upon pistons by one cubic inch hydrogen . . .	287,	340,	280.
Pressure remaining upon pistons 0·2 sec. after complete explosion per sq. inch.	35,	39,	40.
Pressure per piston . .	245,	195,	140.
Mean pressure upon piston . . .	266,	207,	210.

The best mixture with 1 cubic inch of hydrogen only gives a pressure of 267 lbs. available for 0·2 second, so that its capacity for producing power, compared with Glasgow and Oldham gas, is as 267 is to 665 and 640 respectively. To produce equal power with Glasgow gas nearly two-and a-half times its volume of hydrogen is required. The idea is very prevalent among inventors that if pure hydrogen and air could be used, greater power and economy would be obtained; these experiments prove the fallacy of the notion. Hydrogen is the very worst gas which could be used in the cylinder of a gas engine, it is useful in conferring inflammability upon dilute mixtures of other gases, but when present in large quantity in coal gas it diminishes its value per cubic foot for power.

PRESSURES PRODUCED IF NO LOSS OR SUPPRESSION OF HEAT EXISTED.

From the fact already mentioned in the last chapter, that the theoretical temperatures of combustion are never attained in reality, it will naturally be expected that the pressures produced by explosions in closed vessels will also fall short of theory. This is found to be the case. It has been observed by every experimenter upon the subject, beginning with Hirn in 1861, who determined the pressures produced by the explosion of coal gas and air, and hydrogen and air. He used two explosion vessels of 3 and 36 litres capacity; they were copper cylinders with diameters equal to their length. He used a Bourdon spring manometer to register the pressure. He states that:

(1) With 10 per cent. hydrogen introduced the results were: according to experiment, 3·25 atmospheres; according to calculation, 5·8 atmospheres.

(2) With 20 per cent. of hydrogen, the results were: according to experiment, 7 atmospheres, which is very much below the calculation.

(3) With 10 per cent. of lighting gas introduced the results were: according to experiment, 5 atmospheres, *i.e.* much more than with the introduction of an equal volume of pure hydrogen.

He notices especially the low pressure produced by hydrogen as compared with lighting gases, but observes truly that this should not excite surprise—although the heat value of hydrogen is great, yet it is so when compared with equal weights of other substances—and that coal gas being four or five times as heavy as hydrogen, quantity is balanced against quality; therefore volume for volume it gives out more heat.

He considers that there is no difficulty in explaining the very considerable difference found between calculation and experiment, as the metal sides are at so low a temperature compared with the explosion, that the heat is rapidly conducted away, and the attainment of the highest temperature is impossible. Bunsen, in his experiments, observed the same difference, and so later did Mallard and Le Chatelier. The author's experiments fully confirm the accuracy of those observers. In no case, whether with weak or strong mixtures of coal gas and air, or hydrogen and air, is the pressure produced which should follow the complete evolution of heat.

Thus, with hydrogen mixtures (*Clerk's experiments*):

	Per sq. in.	
1 vol. H 6 vols. air gives by experiment	41 lbs.	above atmosphere.
The calculated pressure is	88·3	,, ,,
1 vol. H 4 vols. air experiment gives	68	,, ,,
Calculated pressure is	124	,, ,,
2 vols. H 5 vols. air experiment gives	80	,, ,,
Calculated pressure is	176	,, ,,

Without exception the actual pressure falls far short of the calculated pressure; in some manner the heat is suppressed or lost. That the difference cannot altogether be accounted for by loss of heat is easily proved; the fall of pressure is so slow from the maximum that it is impossible that any considerable proportion of heat can be lost in the short time of explosion. If so large a proportion were lost on the rising curve, it could not fail to show upon the falling curve; it would fall in fact as quickly as it rose. Again, the increase of pressure would be less in a small than in a large vessel, as the small vessel exposes the larger surface proportionally to the gas present. It is found that this is not so. Bunsen used a vessel of a few cubic centimetres capacity, and got with carbonic oxide and oxygen true explosive mixture 10·2 atmospheres maximum pressure; Berthelot with a vessel 4000 cb. c. capacity got 10·1 atmospheres; with hydrogen true explosive mixture Bunsen 9·5 atmospheres, Berthelot, 9·9 atmospheres. All the difference, therefore, cannot be accounted for by loss before complete explosion.

Mixtures of air and coal gas give similar results.

The following are the observed and calculated pressures for Oldham coal gas. (*Clerk's experiments.*)

	Per sq. in.	
1 vol. gas 14 vols. air, experiment gives	40 lbs. above atmosphere	
Calculated pressure is	89·5	,, ,,
1 vol. gas 13 vols. air, experiment gives	51·5	,, ,,
Calculated pressure is	96	,, ,,
1 vol. gas 12 vols. air, experiment gives	60	,, ,,
Calculated pressure is	103	,, ,,
1 vol. gas 11 vols. air, experiment gives	61	,, ,,
Calculated pressure is	112	,, ,,
1 vol. gas 9 vols. air, experiment gives	78	,, ,,
Calculated pressure is	134	,, ,,
1 vol. gas 7 vols. air, experiment gives	87	,, ,,
Calculated pressure is	168	,, ,,
1 vol. gas 6 vols. air, experiment gives	90	,, ,,
Calculated pressure is	192	,, ,,

The results with Glasgow gas are so similar that it is unnecessary to give a table; in no case does the maximum pressure account for much more than one-half of the total heat present. As all of the deficit cannot have disappeared previous to complete explosion, it follows that the gases are still burning on the falling curve, that is, the falling curve does not truly represent the rate of cooling of air heated to the maximum temperature, because heat is being continually added by the continued combustion of the mixture. This will be fully proved by a study of the curves.

It may, however, be taken as completely proved by the complete accord of all physicists who have experimented on the subject, that for some reason nearly one-half of the heat present as inflammable gas in any explosive mixture, true or dilute, is kept back and prevented from causing the increase of pressure to be expected from it. Although differences of opinion exist on the cause, all are agreed on the fact; they also agree in considering that inflammation is complete when the highest pressure is attained.

Temperatures of Explosion.

With a mass of any perfect gas confined in a closed vessel the absolute temperatures and pressures are always proportional; double temperature means double pressure. Temperatures T, t (absolute), pressures corresponding P, p; then $\frac{T}{t} = \frac{P}{p}$ (Charles's law). If explosive mixtures behaved as perfect gases, the pressure before explosion and temperature being known, the pressure of explosion at once gives the corresponding temperature. It has been shown at page 82 that explosive mixtures do not fulfil this condition, but change in volume from chemical causes quite apart from physical ones. It follows, therefore, that these changes must be known before the temperature of the explosion can be calculated from the pressure. In the cases of hydrogen and carbonic oxide true explosive mixtures with oxygen, a contraction of volume is the result of combination. It comes to the same thing as if a portion of the perfect gas in the closed vessel was lost during heating; the temperature then could not be known at the higher pressure unless the volume lost is also known.

Suppose one-third of the volume to disappear, upon cooling to the original temperature, the pressure would be reduced to two-thirds of the original pressure, and this fraction of the original pressure must be taken as $p_1 = 10$. As both steam and carbonic acid at temperatures high enough to make them perfectly gaseous occupy two-thirds of the volume of their free constituents, it follows that p_1 must be taken as $\frac{2}{3}\,p$, wherever the temperatures are such that combination is complete. But here another difficulty occurs. Bunsen found that hydrogen and oxygen in true explosive mixtures gave an explosion pressure of 9·5 atmospheres. The calculated pressure for complete combustion, and allowing for chemical contraction is 21·3 atmospheres. It is evident enough that complete combustion has not occurred, but it is difficult to say what fraction remains uncombined. Yet unless the fraction in combination be known the contraction cannot be known, and therefore the temperature corresponding to the pressure cannot be known.

Berthelot has pointed out that in a case of this kind the true temperature cannot be calculated, but it may be shown to lie between two extreme assumptions, both of which are erroneous.

(1) Temperature calculated on assumption of no contraction.

(2) Temperature calculated on assumption of the complete contraction.

Let the two temperatures be (1) T^1 and (2) T.

	T^1	T
2 vols. H, 1 vol. O, explosion pressure (absolute) 9·9 atmospheres	2449° C.	3809° C.
2 vols. CO, 1 vol. O, explosion pressure (absolute) 10·8 atmospheres	2612° C.	4140° C.

The lower temperature could only be true if no combination whatever had occurred, which is impossible, as then no heat at all could be evolved; the higher temperature could only be true if complete combination, and therefore complete contraction, occurred. The truth is somewhere between these numbers.

When the explosive mixture is dilute, the limits of possible error are narrower, because the possible proportion of contraction is less; with hydrogen and air mixture in proportion for complete combination, 2 volumes of hydrogen require 5 volumes of air. The greatest possible contraction of the 7 volumes is therefore 1 volume. If all the hydrogen burned to steam, the 7 volumes contract to 6 volumes. With more dilute mixtures the proportion diminishes.

With a mixture containing $\frac{1}{5}$ of its volume hydrogen, 10

Explosion in a Closed Vessel

volumes can only suffer contraction to 9 volumes. With ½ volume hydrogen, 14 volumes can contract to 13 volumes.

The limits of maximum temperatures for those mixtures are as follows (*Clerk*):

	T'	T
1 vol. H, 6 vols. air, explosion pressure (absolute), 55.7 lbs. per sq. in.	826° C.	909° C.
1 vol. H, 4 vols. air, explosion pressure (absolute), 82.7 lbs. per sq. in.	1353° C.	1539° C.
2 vols. H, 5 vols. air, explosion pressure (absolute), 94.7 lbs. per sq. in.	1615° C.	1929° C.

The possible error is here much less than with true explosive mixtures; coal gas is of such a composition that some of its constituents expand upon decomposition previous to burning, and so to some extent balance the contraction produced by the burning of the others. The possible error is therefore still further reduced. The composition of Manchester coal gas as determined by Bunsen and Roscoe is as below. The oxygen required for the complete combustion of each constituent is also given, and the volumes of products formed.

ANALYSIS OF MANCHESTER COAL GAS. (*Bunsen and Roscoe.*)

	vols.	Amount required for complete combustion vols. O	Products vols.
Hydrogen, H	45.58	22.79	45.58, H_2O
Marsh gas, CH_4	34.9	69.8	104.7, CO_2 & H_2O
Carbonic oxide, CO	6.64	3.32	6.64, CO_2
Ethylene, C_2H_4	4.08	12.24	16.32, CO_2 & H_2O
Tetrylene, C_4H_8	2.38	14.28	19.04, CO_2 & H_2O
Sulphuretted hydrogen, H_2S	0.29	0.43	0.58, H_2O & SO_2
Nitrogen, N	2.46	—	2.46
Carbonic acid, CO_2	3.67	—	3.67
Total	100.00	122.86 O	198.99, CO_2 H_2O & SO_2

When burned in oxygen 100 volumes of this sample of gas require 122.86 volumes of oxygen, total mixture 222.86 volumes; the products of the combustion measure 198.99 volumes. Calculating to percentage, 100 volumes of the mixture will contract to 89.4

volumes of the products. As 100 volumes of the mixture will contain 55·1 volumes of oxygen, it follows that if air be used, four times that volume of nitrogen will be associated with it, that is, 55·1 × 4 = 220·4. The strongest possible explosive mixture of this coal gas with air containing 100 volumes of the true explosive mixture will be 320·4 volumes, and it will contract upon complete combustion to 309·8 volumes.

One volume of this gas requires 6·14 volumes air for complete combustion, and 100 volumes of the mixture contract to 96·6 volumes of products and diluent. A contraction of 3·4 per cent. Dilution still further diminishes the change; thus a mixture, 1 volume gas 13·28 volumes air, will have only half that contraction, or 1·7 per cent.

From these figures it is evident that the limits of possible error in calculating temperature from pressure of explosion does not exceed, in the worst case, with coal gas and air 3·4 per cent., and in weaker mixtures half that number. The fact that the whole heat is not evolved at the explosion pressure, and that therefore the whole contraction does not occur then, further reduces the error. It is then nearly correct to calculate temperature from pressure without deduction for contraction. This has been done for Glasgow gas and for the Oldham gas experiments by the author.

Explosion in a Closed Vessel. (*Clerk.*)
Mixtures of air and Glasgow coal gas.

Temp. before explosion . . . . . . 18° C.
Pressure before explosion . . . . . atmos. 14·7 lbs.

Mixture		Max. press. above atmos. in pounds per sq. in.	Temp. of explosion calculated from observed pressure
Gas.	Air.		
1 vol.	13 vols.	52	1047° C.
1 vol.	11 vols.	63	1265° C.
1 vol.	9 vols.	69	1384° C.
1 vol.	7 vols.	89	1780° C.
1 vol.	5 vols.	96	1918° C.

Mixtures of air and Oldham coal gas.

Temp. before explosion . . . . . 17° C.

Mixture		Max. press. above atmos. in pounds per sq. in.	Temp. of explosion calculated from observed pressure	Theoretical temp. of explosion if all heat were evolved
Gas.	Air.			
1 vol.	14 vols.	40	806° C.	1786° C.
1 vol.	13 vols.	51·5	1033° C.	1912° C.
1 vol.	12 vols.	60	1202° C.	2058° C.
1 vol.	11 vols.	61	1220° C.	2228° C.
1 vol.	9 vols.	78	1537° C.	2670° C.
1 vol.	7 vols.	87	1733° C.	3334° C.
1 vol.	6 vols.	90	1742° C.	3808° C.
1 vol.	5 vols.	91	1812° C.	
1 vol.	4 vols.	80	1595° C.	

Those temperatures calculated from maximum pressure, although not quite true are very nearly so, whatever be the theory adopted to explain the great deficit of pressure. It does not follow, however, that they are the highest temperatures existing at the moment of explosion; they are merely averages. The existence of such an intensely heated mass of gas in a cold cylinder causes intense currents, so that the portion in close contact with the cold walls will be colder than that existing at the centre. There will be a hot nucleus of considerably higher temperature than that outside, but whatever that temperature may be, the increase of pressure gives a true average. It may be taken, then, that coal gas mixtures with air give upon explosion temperatures ranging from 800° C to nearly 2000° C., depending on the dilution of the mixture. The more dilute the mixture the lower the maximum temperature; increase of gas increases maximum temperature at the same time as it increases inflammability.

The author has made explosion experiments in the same vessel with mixtures previously compressed, and finds that the pressures produced with any given mixture are proportional to the pressure before ignition, that is, with a mixture of constant composition, double the pressure before explosion, keeping temperature constant at 18° C., doubles the pressure of explosion. The experiments are laborious, and they are not yet complete for publication, but the general principles already developed are true for compressed mixtures also.

Efficiency of Gas in Explosive Mixtures.

Rankine defines available heat as follows :

'The available heat of combustion of one pound of a given sort of fuel is that part of the total heat of combustion which is communicated to the body to heat which the fuel is burned ; and the efficiency of a given furnace, for a given sort of fuel, is the proportion which the available heat bears to the total heat.'

The gas engine contains furnace and motor cylinder in one ; nevertheless the efficiency of the working fluid is quite as distinct from the furnace efficiency as in the steam engine. Rankine's definition is quite true for the gas engine.

The fuel being gas, the working fluid consists of air and its fuel and their combinations ; the available heat is that part of the heat of combustion which serves to raise the temperature of the working fluid ; the part which flows into it to make up for loss to the cold cylinder walls cannot be considered available. To be truly available it must either increase temperature, or keep it from falling by expansion. The heat flowing through the cylinder walls is a furnace loss, incident to the explosion method of heating.

The experiments upon explosion in a closed vessel provide data for determining the furnace efficiency as distinguished from that of the working fluid. The proportion of heat flowing from an explosion to the walls in unit time will depend upon the surface of the walls for any given volume. The smaller the cooling surface in proportion to volume of heated gases, the slower will be the rate of cooling. Therefore to be applicable to any engine, the explosion vessel in which the experiments are made should have the same capacity and surface as the explosion space of the engine.

The author's experiments are therefore only strictly applicable to engines with cylinders similar to his explosion vessel. Within certain limits, however, the error introduced by applying them to other engines is inconsiderable.

Assuming the stroke of a gas engine (after explosion) to take 0·2 second, this may be taken as the time during which the pressure of explosion must last if it is to be utilised by the engine. In a closed vessel the pressure falls considerably in 0·2 second, the average pressure may be taken as nearly indicating the available pressure during that time. The heat necessary to produce that pressure is the available heat; and its proportion to the total heat which the gas present in the mixture can evolve is the efficiency of the gas in that explosive mixture.

With Oldham gas the best mixture is (table, p. 103) 1 volume gas 12 volumes air; the average pressure during the first fifth of a second is 51 lbs. per square inch above atmosphere. If all the heat present heated the air, the pressure should be 103 lbs. effective, so that the efficiency of the heating method is $\frac{51}{103} = 0·49$.

The strongest mixture which still contains oxygen in excess is 1 volume gas 7 volumes air, the average available pressure is 67 lbs. per square inch (all heat evolved would give 168 lbs.), the efficiency is $\frac{67}{168} = 0·40$ nearly.

Calculated in this way the efficiency values for Oldham gas mixtures are:

Prop. of Oldham gas in mixture	$\frac{1}{15}$	$\frac{1}{14}$	$\frac{1}{13}$	$\frac{1}{12}$	$\frac{1}{10}$	$\frac{1}{8}$	$\frac{1}{7}$
Heating efficiency	0·40	0·48	0·50	0·43	0·46	0·40	0·37

The furnace efficiency plainly diminishes with increased richness of the mixture in gas.

TIME OF EXPLOSION IN CLOSED VESSELS.

The rates of the propagation of flame in explosive mixtures given in tables, pages 86 and 87, are true only where the inflamed portion is free to expand without projecting itself into the unignited portion. They are the rates proper for constant pressure.

Where the volume is constant, in a closed vessel, the part first inflamed instantly expands and so projects the flame surface into the mass, compressing what remains into smaller space.

To the rate of inflammation at constant pressure are added the projection of the flame into the mass by its expansion and also the increased rate of propagation in the unignited portion by the heating due to its compression by portion first inflamed.

It follows that the rate continually increases, as the inflammation proceeds until it fills the vessel.

This is evident from all the explosion curves. The pressure rises slowly at first, then with ever increasing rate till the explosion is complete; thus the explosion curve for hydrogen mixture with air $\left(\frac{2}{7} H\right)$, shows an increase of 17 pounds in the first 0·005 second, the maximum pressure of 80 pounds being attained in the next 0·005 second. With the weaker mixtures the same thing occurs, rise of pressure, slow at first, then more rapid, and in some cases becoming slow again before maximum pressure. The time taken to get maximum pressure varies much with the circumstances attending the beginning of the ignition. If a considerable mass be ignited at once, by a long and powerful spark, or by a large flame, the ignition of the weakest mixture may be made almost indefinitely rapid. Something very like Berthelot's explosive wave may result. This is due to the great mechanical disturbance caused by the rapid expansion of the portion first ignited; the smaller that portion is the more gently does the flame spread. A small separate chamber connected with the main vessel, if filled with explosive mixture and ignited, will project a rush of flame into the main vessel and cause almost instantaneous ignition. The shape of the vessel, too, has a great effect upon the rate. Where it is cylindrical and large in diameter proportional to its axial length, ignition is extremely rapid, the flame is confined at starting, and is rapidly deflected by the cylinder ends, and so shoots through the whole mass.

By so arranging the explosion space of a gas engine that some mechanical disturbance is permitted, it is easy to get any required rate of ignition even with the weakest mixtures.

The maximum pressure is not increased by rapid ignition.

Starting the ignition from a small spark, the time taken to ignite increases with the volume of the vessel.

Berthelot has experimented upon this point with explosion vessels of three capacities, 300 cubic centimetres, 1500 cubic centimetres, and 4000 cubic centimetres. He finds time of explosion (he also takes maximum pressure to indicate complete explosion) of mixture 2 vols. H, 1 vol. O, and 2 vols. N, in 300 cubic centimetre vessel, 0·0026 second; and in 4000 cubic centimetre vessel, 0·0068 second.

With mixture of carbonic oxide and oxygen, 2 vols. CO, 1 vol. O, smaller vessel, 0·0128 second; larger vessel, 0·0155 second. Mixtures with air were much slower. The conclusion then is obvious, that in large engines the time of explosion will be longer than in small ones.

CHAPTER VII.

THE GAS ENGINES OF THE DIFFERENT TYPES IN PRACTICE.

HAVING now studied the theoretic efficiency of the different kinds of engine and the mechanism of the heating method—that is the properties of gaseous explosions—the way is clear for the study of the results obtained from the engines in practice.

It is quite evident that no practicable engine can give an efficiency at all approaching theory from the use of gaseous explosions; the temperatures and therefore pressures produced fall far short of that due to the complete evolution of the heat present in the mixture as combustible gas. All the heat of the gas does not go to increase the temperature of the working fluid; a large proportion of it is rendered latent in some way when the maximum temperature is attained.

The appearance of the diagrams from the explosion of mixtures commonly used in gas engines, shows at first a very rapid increase of pressure and temperature, which terminates abruptly and is immediately succeeded by a fall which is relatively a slow one.

It was formerly supposed that the completion of the explosion was coincident with the completion of the combustion, and therefore of the evolution of heat. This, however, was shown by Bunsen and those who have followed him, to be untrue; although the temperature ceases to rise, and fall sets in, the gas present has in few explosions been more than half-burned at the moment of maximum temperature. The causes which suppress the heat of the explosion and prevent it from being evolved at once are complex and have occasioned different explanations which will be fully discussed in a subsequent chapter. Meantime, it is sufficient to recognise the fact and to understand its bearing upon the economy of gas engines.

It is a phenomenon common to all gas engines which have ever been constructed, whether using compression previous to ignition or not. The heat so suppressed appears when cooling sets in, and consequently explosive mixtures cool more slowly in appearance than would a mass of air heated to similar temperatures and exposed to similarly cold enclosing walls.

In many gas engines the indicator diagrams are apparently almost perfect, that is, the lines of falling temperatures are almost true adiabatics. So far as the diagram yields information, the gases in expanding are losing no heat whatever to the cylinder, but the temperature is falling apparently only by work done upon the piston. This supposition is known to be untrue, because the gases are at a temperature often as high as the hottest of blast furnaces, and the walls enclosing are at most at the boiling point of water. It is the suppressed heat which is being evolved during fall of temperature which sustains the temperature and makes the diagram appear as if no loss or but little was going on. An actual engine therefore may give a diagram which is the exact theoretical one, and yet the efficiency of the engine be much below theory. The author's experiments upon explosive mixtures were undertaken to get the data necessary for the interpretation of the diagram, and the rising and falling curves, showing times of rise and fall of pressure, give the efficiency of coal gas in the different mixtures, apart altogether from theoretic considerations. Whatever the opinions held regarding the cause or causes of the suppression of heat, the experiments with carefully proportioned explosive mixtures, at known temperatures and pressures, determine absolutely the capability of gas for producing pressure and for sustaining it under cooling.

As the efficiency may be very different from that shown by the indicator, it is advisable to distinguish between the real and apparent efficiency. Call the one *apparent indicated* efficiency, and the other *actual indicated* efficiency.

The apparent indicated efficiency, when multiplied by the efficiency of the gas in the particular mixture used, will give the actual indicated efficiency. For instance, if the diagram gave the efficiency of an engine as 0·29 and the efficiency of the mixture was 0·48, then the actual indicated efficiency is 0·29 × 0·48 = 0·11. That is, only

0.48 of the gas present when the diagram is taken really acts in producing elevation of temperature; the remaining 0.52 is suppressed and keeps up the temperature, which would otherwise fall by cooling. The diagram alone can never tell accurately the losses which are taking place unless the heat is all evolved at once and appears in temperature; then, but not till then, will the lines traced by the indicator tell the loss of heat. Some previous writers have misinterpreted their indicator diagrams through neglect of this fact.

Some others, notably Dr. Slaby, of Berlin, have assumed that the phenomenon of retarded combustion is produced by invention and occurs only in the Otto engine. This is a mistake. All engines using explosion necessarily exhibit it ; in fact, as it is an accompaniment of all explosions, it is impossible to make an engine in which it is avoided.

In the following examination of the performances of the various engines in practice the importance of the phenomenon will appear.

Type 1.—The most important engines of this type which have yet been in public use are those of Lenoir, Hugon, and Bisschoff. Many others have been made and sold in some numbers, but as these three present fully all the peculiarities of the type, it would only waste time to describe the varied mechanical details constituting the sole novel points in the others.

Lenoir Engine.

The Lenoir engine as made differs considerably from that described in his specifications. As a rule of almost general application, specifications are untrustworthy as accurate descriptions of working machines ; the author has been careful to describe no engine which he has not examined.

Fig. 22 is a section of the cylinder of a half-horse power Lenoir engine. The engine in the Patent Office Museum, South Kensington, is well made and in external appearance closely resembles an ordinary high-pressure steam engine.

The cylinder is $5\frac{1}{2}$ inches diameter, and the stroke is $8\frac{1}{2}$ inches. Its cylinder is provided with two valves ; both are slides, working between the cylinder faces and covers, which are held down to the slides by adjusting screws. One valve controls the discharge of the products of combustion, the other, the admission and mixing of the inflammable gas and air. The ignition is effected by the electric spark. The working cycle of the engine is as follows :

When the piston is at the end of its stroke, the gas and air admission valve is open ; the main port in it opens to the atmosphere,

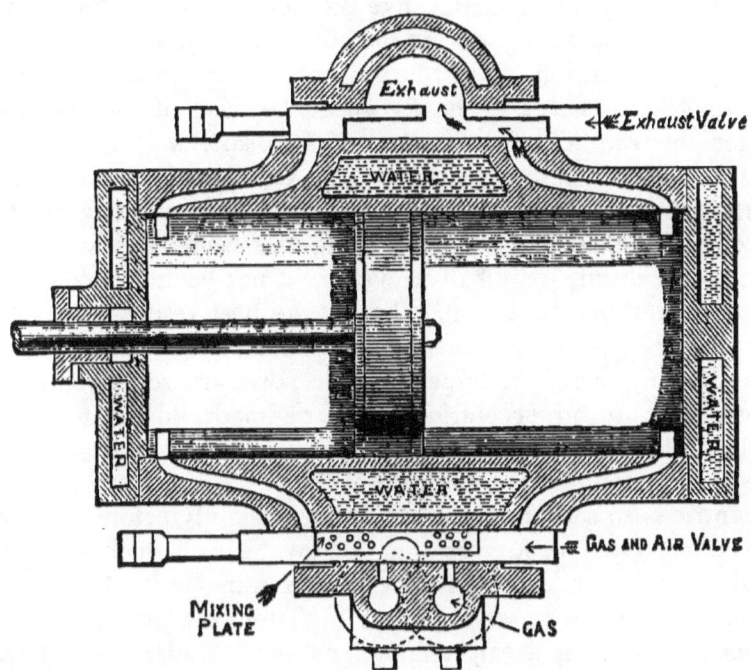

FIG. 22.—Lenoir Engine Cylinder (sectional plan).

while a smaller port leads from the main port to the gas supply. The forward movement of the piston draws into the cylinder the air and the gas, which mix as they enter the main valve port and the engine admission port. At about half-stroke the supply of mixed gases is cut off, so that the cylinder is completely closed off from the atmosphere and from the gas supply ; an electric spark now passed into the explosive mixture from a battery and induction coils causes explosion and the pressure rapidly rises. The piston is thereby pushed on its stroke during the portion remaining to be completed; at the end of the stroke the pressure has fallen by expansion doing work and by the cooling action of the cylinder walls, to nearly atmosphere again; the exhaust valve opens and during the return stroke the products of combustion are expelled preparatory to taking in a fresh charge upon the next working stroke. The same operation is repeated upon the other side of the piston so that the engine is double-acting of a kind. It cannot be considered as truly double-acting, like the steam engine, as the driving pressure is not acting during the whole forward stroke, but only during that portion of it which is not taken up in sucking in the explosive charge. The fly wheel, because of this, is much larger than in a steam engine of corresponding dimensions, and the power is also much less. The valves are both actuated by eccentrics upon the crank shaft. Each slide requires a separate eccentric because the exhaust during the whole stroke and the admission during only half-stroke could not be managed by the single to and fro movement. To get the best result it is evident that the least possible power should be expended in introducing the charge; therefore, large inlet air ports are required, all the larger because an eccentric cannot be made, alone, to give a sudden cut-off. To prevent throttling as the ports approach the closing points, the total opening must be considerable. The eccentric is so set that the port is open slightly before the crank has crossed the centre, so that it may be well open when the charge begins to enter. In fact it has some lead like a steam slide, and for the same purpose. The exhaust valve is set precisely as in the steam engine, and is of similar construction, except that it is not enclosed in a case, but it is held against the cylinder face by a cover and screws. Fig. 22 is a sectional plan of the cylinder, showing the valves and ports; fig. 23 is a transverse vertical section of the cylinder, showing the valves and valve covers with gas, air and exhaust ports. The arrows indicate the direction of the gas and air flow, while mixing and entering the cylinder, also the exhaust path. As the air port opens to the cylinder slightly before the piston has completed its stroke, and a slight pressure may yet remain in the cylinder, the gas port does not open till a little later. The gas and air do not open quite simultaneously, although nearly so; neither do they close quite together. There is one gas admission port in the slide leading into the main port; in the cover there are two ports between which the slide port passes, taking gas from either, as is required, for the end of the cylinder which is receiving the charge. The main valve port opens on the upper side to the air, and is covered by a perforated plate and a light metal case furnished with a throttle valve; the brass plate perforated is carried downwards and covers the gas port, so that the gas entering from the supply pipe is not permitted to flow at once into

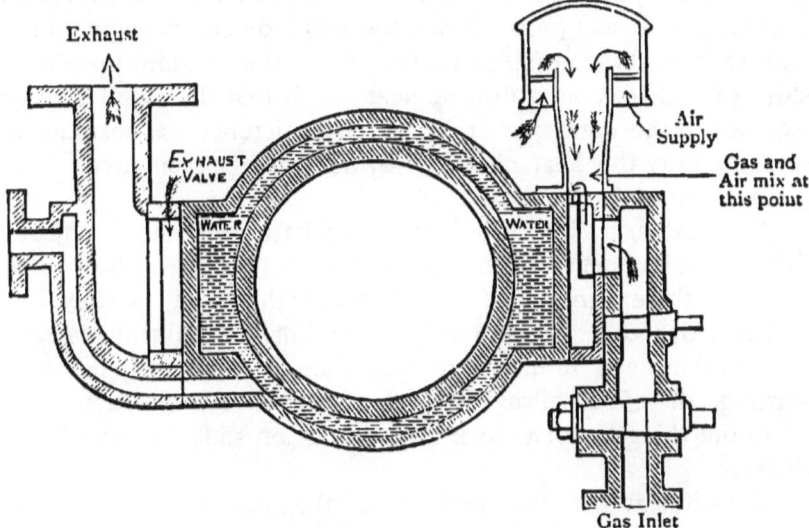

FIG. 23.—Lenoir Engine Cylinder (transverse section).

the main port, but must first pass up through the perforations and mix with the air which is going down through those adjoining. The mixing arrangement is somewhat imperfect, and is exceedingly sensitive to change of speed in the engine. The throttle-valve is intended to increase or diminish the supply of air; by closing it slightly, the suction upon the gas port is increased, and so the proportion of the mixture altered. By opening it, the air has freer access to the cylinder, the pressure is not reduced so much, and therefore the gas is diminished. The mixing, however, is too irregular; as the gas streams are not projected separately into the incoming air stream, the gas flows too much in mass into the air in mass. The igniting points were invariably placed at the upper part of the cylinder in the cylinder covers. The cylinder and covers are waterjacketed, the water is kept continuously flowing through, so that the temperature may not become so high as to injure the cylinder. This is a most necessary precaution in any gas engine of even moderate power; the effect of neglect in a Lenoir engine is very soon observed in complete cutting up of the cylinder; it speedily becomes red-hot if allowed to run without water. Indeed, even with an adequate water supply, the larger engines gave great trouble; although the cylinder could be kept cool the piston could not. It was proportioned too much on steam engine lines, and when working at full power the incessant explosions upon both sides caused so rapid a flow of heat into it that the small surface exposed to the water jacket by the circumference was insufficient to carry away the heat absorbed by the whole piston area. The pistons often became red-hot.

The exhaust slide also had rather hard work, and required delicate adjustment, as the exhaust gases were very hot, often 800° C.; the expansion of the slide was therefore considerable, and in order to be pressure tight when hot, the adjusting screws had to be kept rather easy when cold. The engine when starting, therefore, always leaked a little at the exhaust valve. The same thing happened with the admission slide, but to a lesser degree.

Notwithstanding the large area of the admission port and the lead given to the admission valve, the closing motion was too slow to prevent throttling; accordingly the pressure fell somewhat below atmosphere, while the valve was cutting off preparatory to explosion. After cutting off a slight delay occurred between the passing of the spark and the commencement of the explosion; the explosion itself took some time to complete; it was by no means instantaneous; the diagram produced was consequently imperfect. In addition to all this, the piston being so hot, heated the charge while it was entering, and so occasioned further loss.

The lubricating arrangements also were primitive. The steam engine requiring but little care in lubricating, the gas engine was not supposed to require more; and the ordinary lubricating cock was deemed sufficient. All these sources of loss, inevitable in a first attempt, made the engine comparatively inefficient. Notwithstanding all its defects, the Lenoir engine at the time of its production was the best the world had yet seen, and in careful hands it did good work and created a widespread interest. The engine in South Kensington Museum, under the skilful care of Mr. S. Ford, worked for many years supplying all the power required for the repair department of the Patent Office Museum. It runs with perfect smoothness, nothing whatever in its action would enable one standing beside it to imagine for a moment that the motive power was explosive. The popular notion of an explosion is always associated with the idea of a great noise. This, of course, physicists have always known to be a fallacy, as no explosion makes noise unless it has access to the atmosphere. An explosion in a closed vessel makes no sound unless the vessel bursts. In a gas engine it is only necessary to see that the explosion is not too rapid, but that time is allowed for the slack of the connecting rod and crank connections to take up. The explosions used by Lenoir were seldom more rapid in rise of pressure than is common with all steam engines. A 1-horse Lenoir engine inspected lately by the author at Petworth House, Petworth, had been at work for the past twenty years pumping water for the town and is still at work. It works with smoothness and is altogether more silent in its action than most modern gas engines. The author finds that many Lenoir engines are still at work after twenty years' continuous use, notably two 1-horse power engines at the Brewery of Messrs. Trueman, Hanbury and Buxton, London, and one 1-horse power at the establishment of Messrs. Day, Son and Hewitt, Dorset Street, London, all doing hard work with great regularity.

Diagrams and Gas consumption.—Prof. Tresca of Paris has made experiments with a ½-horse Lenoir engine, and found that it consumed 95 cb. ft. of Paris gas per indicated horse power per hour. The diagrams from so small an engine hardly do justice to the method, and as it is desirable to compare the engine with modern engines using similar volumes of charge the author has taken a diagram from a paper by Mr. Slade, published in the Journal of the Franklin Institute, Philadelphia. The engine had a cylinder of eight inches diameter and sixteen stroke. The explosion space corresponds closely to that of the author's experimental explosion vessel.

The diagram, fig. 24, at once shows the truth of the preceding discussion of the action of the engine.

AB is the atmospheric line, traced upon the indicator card by the pencil before opening the indicator cock to communicate with the interior of the cylinder; it is the neutral position of the indicator piston while the pressure on both sides of it is at atmosphere; any pressure from within the cylinder pushes up the piston and therefore the indicator pencil. Pressure above atmosphere is registered by lines above that line, pressure below atmosphere

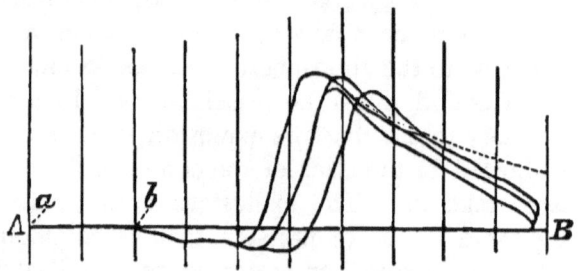

Diagram at 50 revolutions, cylinder 8½ inches diameter, 16¼ inches stroke.

FIG. 24.—Lenoir Engine Diagram.

is registered by lines below that line. The card shows three distinct tracings, each corresponding to one stroke of the engine: admission of the charge, explosion, expansion and return expelling the products of the combustion. If the cycle is carried out in a mechanically perfect manner the admission of the charge should be accomplished without loss by throttling. This is not so. From the point a to the point b the valve is open enough to give free access to the cylinder, and accordingly the pressure within the cylinder is not appreciably lower than that without; but here the valve begins to contract its opening at the very moment that the piston is moving most rapidly, the pressure falls and is a couple of pounds per square inch below atmosphere when it closes. When closed, the spark does not at once take effect, so that the pressure has become 11 lbs. per sq. in. total before the ignition begins to cause a rise. Then the ignition itself takes some time to be completed, here about $\tfrac{2}{15}$ second; the piston has, therefore, moved through a further one-and-a-half-tenth of its stroke and the heat given by the explosion is not added at strictly constant volume, as required by theory. Apart altogether from loss of heat to the cylinder walls, this diagram is mechanically imperfect. The valve arrangements should be such that no loss is incurred in charging and that the explosion follows so rapidly that the pressure in the cylinder has no time to fall by expansion, after closing the admission; the explosion, indeed, should at once follow the cut-off. In the best of the three lines the pressure has

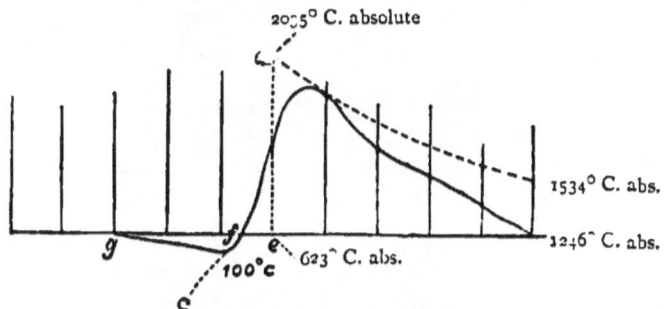

FIG. 25.—Lenoir Engine Diagram.

fallen to nearly 11 lbs. total, and the maximum pressure of the explosion is 48 lbs. per square inch total. The average of the three lines gives a pressure divided over the whole stroke of only 8·3 lbs. per square inch, which, assuming the diagram from the other end of the cylinder to give similar results, gives a total of 2 indicated horse power at 50 revolutions per minute. This is an exceedingly poor result for so large an engine. The apparent indicated efficiency is much below that of a theoretical diagram using the same expansion. Fig. 25 shows in dotted lines a diagram which will have the same efficiency as the actual diagram (best of the three lines). If the temperature of the entering charge has been raised to 100° C., as stated by Mr. Slade, then the point c upon the diagram corresponds to that temperature; the point d will correspond to a temperature of 2035° C. absolute, as the volume has increased from 0·4 to 0·5 and the pressure from 11 lbs. to 14·7 lbs. per square inch total at the point *e*. The area of the part of the explosion curve *def* may be taken as equal to the part of the diagram *cfg* which is resistance due to the valve action; the work done upon the piston by the one part balances the loss by the other; both portions may therefore be neglected, the dotted lines representing the apparent diagram efficiency.

The temperatures for calculating maximum possible efficiency are as follows—they are also marked upon the diagram

T . . . . 2035° absolute.
T^1 . . . . 1534°   ,,
t . . . . 623°   ,,
t^1 . . . . 1246°   ,,

Calculating E from formula (17) p. 57

$$E = 1 - \frac{(T^1 - t^1) + 1\cdot 408 \, (t^1 - t)}{T - t}$$

$$= 1 - \frac{(1534 - 1246) + 1\cdot 408 \, (1246 - 623)}{2035 - 623}$$

$$= 0\cdot 175.$$

The apparent indicated efficiency for the best of the three lines is 0·175. If it were constantly repeated, the actual indicated efficiency may be obtained by multiplying by the efficiency of the gas in the mixture used to get the explosion. The numbers got from explosion in a closed vessel do not quite represent the conditions of loss in a cylinder with a moving piston. In the first case the loss of pressure and temperature is due solely to the cooling effect of the vessel's walls; in the second the moving piston reduces pressure and temperature by expansion, and at the same time increases the surface exposed. The increased surface, however, will not increase the rate of cooling, as the volume is at the same time increased in a greater proportion. It has been already shown that cooling of a heated mass of gas is independent of the pressure, and depends on the ratio of surface to volume.

In the engine the volume of the hot gases becomes doubled by expansion, but the surface exposed does not double; the cylinder surface increases with the volume, but the piston area and cylinder-cover area remain the same, so that the proportion of surface to volume diminishes instead of increasing. The heat lost to the cylinder and piston and cover in the engine will therefore be no greater than that lost to the enclosing wall of the experimental explosion vessel in a similar time. It will indeed be somewhat less, as in the time taken doing work the temperature will fall by heat disappearing as work. With the closed vessel the fall is due solely to cooling, so that the average temperature during the time of exposure is higher. More work is urgently required by careful physicists to get accurate data. At present the approximation to the efficiency of the gas in different mixtures by closed vessel experiments is the best that can be had; it cannot be greatly in error. The efficiencies obtained from the indicator diagram and the author's experiments will be lower than the truth, the more so the greater the expansion. With engines as at present constructed the difference is but small.

The mixture required to give a temperature of 2035° C. absolute is, for Oldham gas, 1 gas 6 air, and the average pressure during 0·3 sec. from complete explosion is 63 lbs. per square inch above atmosphere, nearly. The time taken to expand in the engine after explosion is 0·3 sec. ; the pressure which should be produced by the explosion of this mixture, if all the heat of the gas went to heat the air and products, 192 lbs. per square inch above atmosphere. That is, the difference between 192 and 63 has gone in heat suppressed at the moment of complete explosion and heat lost while exposed to the influence of the vessel walls during the same period as the effective stroke of the engine.

The efficiency of the gas in the mixture is therefore

$$\frac{63}{192} = 0\cdot33 \text{ nearly,}$$

that is, only one-third is really effective in raising temperature. The actual indicated efficiency will, therefore, be only one-third of the apparent. Three times the amount of heat accounted for by the diagram is required to make the gases used in the explosion show the temperatures and curve of the diagram.

Apparent indicated efficiency × efficiency of gas = actual indicated efficiency :

$$0\cdot 175 \times 0\cdot 33 = 0\cdot 058$$

The actual indicated efficiency of the engine is 0·058 or 5·8 per cent. if this diagram be constantly repeated; but as it is the best of the three lines it requires correction. Taking the worst of the three diagrams, fig. 24 shows the temperature as follows : T, 2035° absolute ; T^1, 1697° absolute ; t, 797° ; t^1, 1243° absolute.

The apparent indicated efficiency is $E = 0\cdot 126$.

The actual indicated efficiency is $0\cdot 126 \times 0\cdot 33 = 0\cdot 0495$ or 4·95 per cent. of the total heat given to the engine.

Tresca calculates the heat transformed into work by the Lenoir tested by him as 4 per cent.

The mean of the best and worst of these diagrams is

$$\frac{5\cdot 8 + 4\cdot 95}{2} = 5\cdot 37,$$

which is higher than the result obtained by this distinguished physicist ; but the difference is sufficiently accounted for by the difference in the dimensions of the engines. Tresca's was only halfhorse, Slade's was two horse.

The Lenoir engine used mixtures ranging in composition from 1 gas and 6 vols. air to 1 vol. gas and 12 vols. air, depending upon the amount of work upon the engine; when there was little work the governor was arranged to throttle the gas and so diminish the proportion present. This was a bad plan, as will be explained in the chapter upon governing. But the effect was to make the engine use all grades of ignitable mixtures from the strongest to the weakest. Apart, however, from all intentional arrangements for governing, these engines tended to govern themselves. An increase of speed always causes the proportion of gas in the mixture to diminish, because the resistance of the small gas port to flow increases more rapidly than the larger air port. It follows that if the ports are proportioned to pass certain volumes at a low rate of speed, at a higher rate the proportion is disturbed, the smaller port giving a greater proportional resistance. The effect is seen in all the diagrams, the ignitions become later and later as the mixture diminishes in inflammability, and after attaining a certain dilution, ignition ceases altogether, or becomes too slow to be of any practical use. In the Lenoir type of engine too slow ignition is an unmixed evil, as the theory of the engine requires rapid ignition. In it the loss of efficiency due to valve and igniting arrangements is considerable. The electric ignition is very delicate and troublesome. To overcome the defects of the Lenoir, Hugon introduced his engine, which in some respects was a considerable advance.

HUGON ENGINE.

The Hugon engine, like the Lenoir, exploded the charge drawn into the cylinder by the piston at atmospheric pressure : in it, however, greater expansion and more dilute mixtures were used.

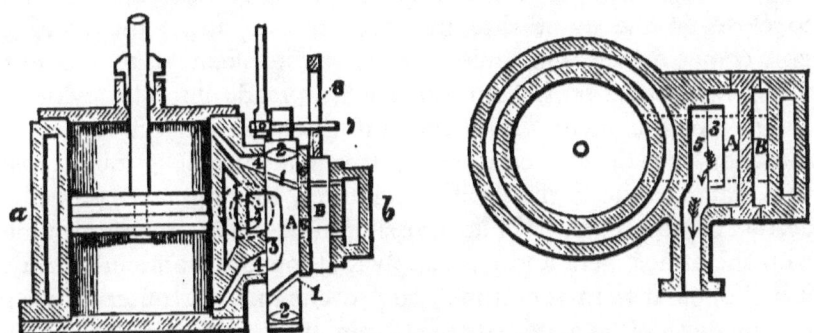

FIG. 26.   FIG. 27.
Hugon Engine Cylinder.

Fig. 26 is a sectional plan showing valves, passages and the cylinder and piston. Fig. 27 is a transverse vertical section through the cylinder at the line ab. The admission of the charge and the expelling of the exhaust are accomplished through the same passage, so that the cylinder has only two ports, as in the steam engine; two valves are used, one working outside the other. The inner valve has five ports, two for admitting the charge, one for exhausting, and two for carrying the igniting flame. The ignition by flame was first accomplished in a workable manner by Hugon, although it had been described in several patents long before his time.

The ports marked 1 1 in the inner slide A are admission, the ports marked 2 2 in the inner slide are igniting, ports; the port 3 is the exhaust passage, alternately communicating with each end of the cylinder by the long ports 4 4 to the exhaust port 5, precisely as in a steam engine. The action of the admission ports is somewhat novel. The object is to secure a rapid opening and cut-off, bringing the igniting flame on immediately after closing the cylinder. The valve is actuated from a cam. When the piston is at the end of its stroke and is moving forward, the valve A is moving in the same direction, the port 3 is allowing the exhaust gases to escape from the other side of the piston, the port 1 is open to the cylinder and is communicating through the port 6 or 6, in the outer slide B, with the air and also with the gas supply. When the piston has taken in sufficient charge, the cam moves the slide A suddenly forward, so causing the port 1 to close on the outer side but not on the inner; the igniting port comes on and the flame burning in it inflames the mixture, filling the engine port, from whence it spreads into the cylinder itself. As the inner valve cuts off when moving in the same direction as it does when opening, it is evident that it must cross back again, to be in the position required to commence opening at the correct time. While crossing, unless the communication with the atmosphere and gas supply is stopped in some other way, it will open at the wrong time; to prevent this, the outer valve B is provided. It is actuated from a pin projecting from the main valve A; this pin 7 works in the slot 8, and while the main valve is moving forward after cutting off, the pin strikes the end of the slot and carries the outer valve with it, causing it to close the port in the cover which it commands. A small plate and spring give friction enough to keep the valve in position till it is moved in the other direction. When the main valve returns, although its ports open on the engine ports, the outer ends are blinded by the outside valve which is not again opened till the main valve has closed. By this ingenious contrivance, a rapid admission and cut-off are secured with one cam and the main and auxiliary slides. The engine from which these details are taken is in South Kensington Museum and is rated at $\frac{1}{2}$-horse power. The valves are arranged to cut off at about one-third stroke.

The cylinder is $8\frac{3}{16}$ diameter and 10 in. stroke. The clearance spaces due to the long ports 4 4, the valve ports open to the cylinder at the moment of explosion, and the space into which the piston does not enter, make up in all a proportion of products of combustion equivalent to nearly thirty per cent. of the entire charge. The effect of this is to cause a considerable difference between the nature of the mixture in the port and that in the cylinder itself, the port mixture being much more inflammable than that in the cylinder. As a consequence the ignition is more rapid with weak mixtures than in the Lenoir. The gas is supplied to the air port in regulated amount by means of a bellows pump worked from an eccentric on the crank shaft; it mixes with the air in passing through the valves and port; the products of combustion are therefore completely expelled from the port, and nothing but pure mixture left to be inflamed by the igniting arrangement. The gas for the internal igniting flame is supplied also from a bellows pump under slight pressure. This flame is extinguished by each explosion, and is relighted when the port opens again to the air by a constant external flame. The action of the exhaust port in the main slide is so evident as to require no other explanation than that afforded by the drawing.

The engine works very smoothly, and is a great improvement upon Lenoir in certainty of action; all the trouble with the battery and coil is very simply avoided. To prevent overheating of the piston, water is injected by means of a tap; it is adjusted so that each suck of the engine drawing in mixture also takes in enough water to keep the piston at a reasonable temperature. In this the engine was successful; it was capable of harder and more continuous work than the Lenoir, and was in every way more certain in its action even with a considerable variation in the composition of the explosive mixture used. The only parts which gave trouble were the bellows pumps controlling the gas supply to cylinder and igniting port; these were made of rubber, and deteriorating after some use gave trouble by leaking and occasional bursting. In some of the engines in use they were replaced by metal pumps and a mixing valve. With these additions the engine in the Patent Office Museum ran for many years.

Diagrams and Gas Consumption.—According to Professor

Tresca, the gas consumed by a Hugon engine of 2-horse power was 85 cubic feet per indicated horse per hour.

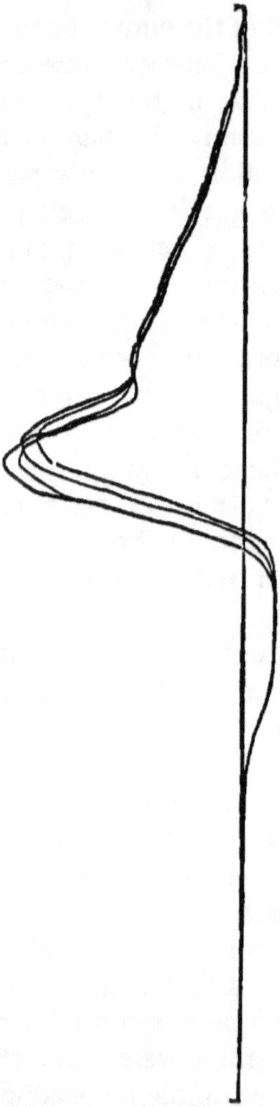

Fig. 28 is a diagram taken from a ½-horse engine by the author. The engine was indicating 0·78 horse power, the average pressure being 3·9 lbs., and the maximum 25 lbs. per sq. in. The card shows considerable delay in explosion after cut-off, notwithstanding the rapid movement of the igniting slide.

BISCHOFF ENGINE.

The consumption of the non-compression type of engine is too high to permit of its use in any but the very smallest machines; accordingly the Lenoir and Hugon engines have long disappeared from the market, and the type survives mainly in the Bischoff, which is specially designed for small powers, mostly under half-horse. It is an exceedingly ingenious little engine, and presents many interesting peculiarities.

Fig. 29 is a side elevation, part in section; fig. 30 a section arranged to explain the valve action. In both figures the similar parts are marked with similar letters. There is no attempt to gain economy by attention to theory; the aim is to get a small workable engine with the least possible complication. In this it is very successful. To avoid the complication of a water-jacket, the cylinder and piston are so arranged that heating is allowable. The engine is upright and very peculiar in appearance, the cylinder has cast on it a number of radiating ribs, which by

FIG. 28. — Diagram from ½-h.p. Hugon Engine; 75 revs. per min. (*Clerk.*) Scale 1 in. = 24 lbs.; 8 3/16″ diam. of cylinder: 10 ins. stroke: 0·78 i.h.p.

contact with the air cause conduction of the heat more rapidly than would otherwise occur. The temperature, however, becomes very high, and provision is made to prevent injury to the piston. It is fitted loosely to the cylinder and has no rings, the connecting

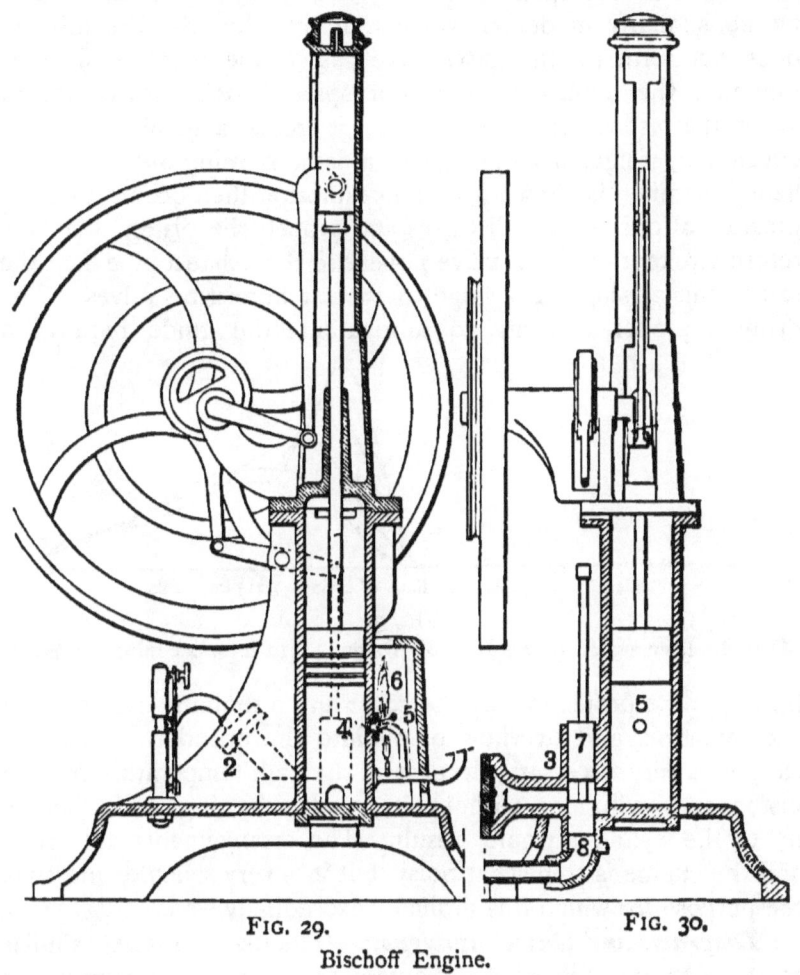

FIG. 29. FIG. 30.
Bischoff Engine.

rod arrangement is seen in the figure (29); it takes the thrust of the explosion in tension, and almost without side pressure upon the guide. Any side pressure upon the guide is quite prevented from reaching the piston, and it consequently is never rubbed against the cylinder. The pressure of the explosion is so slight that the leakage is not serious even without rings. The piston moves up, taking in the charge, the air through the valve 1, fig. 30, which is simply a piece of sheet rubber backed by a thin iron disc. The pressure of the air opens, and the explosion closes it; the valve 2, fig. 29, similarly made but smaller, admits the gas; the mixture does not form till the gases have passed the point 3, fig. 30; therefore the explosion does not spread back to the valves. When the piston gets to the point 4, it crosses a small aperture 5 covered by a light hanging valve; a flame burning outside in the flame chamber is drawn in. The explosion then occurs, and the pressure at once closes all valves and propels the piston. On the return stroke, the piston valve 7 opens to the exhaust pipe 8, at the same time closing the passage to the air admission valves. The cylinder proper requires no lubrication; the guide requires a

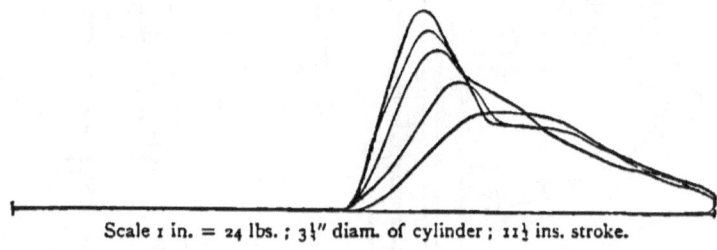

Scale 1 in. = 24 lbs.; 3¼" diam. of cylinder; 11½ ins. stroke.
FIG. 31.
Diagram from 1-man power Bischoff Engine; 112 revs. per min. (*Clerk*.)

little, but the projection of the cover and a draining hole prevent accumulation and overflow of oil into the cylinder. This precaution is very necessary, because of the high temperature of both piston and cylinder; without it speedy charring and choking up of the cylinder would result. The arrangements are crude and the engine is somewhat noisy, but it is very reliable, and suits the purpose for which it is designed exceedingly well.

Diagrams and Gas Consumption.—The diagram is very similar to the Lenoir. Fig. 31 is a diagram taken from a 1-man power engine by the author.

The consumption is, as might be expected, rather higher than Lenoir. According to tests made at the Stockport Exhibition it uses 120 cubic feet per actual horse power per hour.

Type (Ia).

Free Piston Engines.—The very high consumption of gas common to the engines described prevented their extended use, and set inventors to work to produce some method which would give better results. It was very obvious that there was a large loss of heat; the trouble with cylinders and pistons made this abundantly evident. Devices proposed for increasing power by the injection of water spray, and steam, in various ways failed to produce good effect except in aiding lubrication. The inventors of the day seem to have reasoned somewhat in this fashion. The force generated by an explosion of gas and air is an exceedingly evanescent one, a high pressure is produced, but it lasts only for a very short time; if work is to be obtained before loss by cooling absorbs all the heat, it must be done rapidly. The reason why the Lenoir and Hugon engines give so poor a result is a too slow movement of piston after the explosion. Therefore, if a method can be devised permitting greater piston velocity, better economy will be obtained. In this reasoning there was considerable truth. It has been already proved that the shorter the time of contact between the charge after explosion and the enclosing walls, the greater will be the efficiency of the gas in the mixture. But this only holds within certain limits. If the expansion is too rapid before explosion is complete, then a loss instead of a gain will occur; the expansion should not commence till maximum pressure is attained or it will cause a loss of pressure. Indeed, it is quite conceivable that in engines of the Lenoir type, the expansion might be so rapid, relatively to the rate of explosion, that no increase of pressure at all resulted; in which case no power whatever would be obtained. The gain then to be expected arises from rapid expansion after complete explosion. This has been carried out by several inventors by the free piston method. Instead of expending the force of the explosion upon a piston rigidly connected to a crank, the piston is allowed free movement. The explosion launches it against the atmosphere; it acquires considerable velocity, which is expended in compressing the exterior atmosphere, that is, in producing a vacuum in the cylinder. When all the energy of motion is expended, the piston comes to rest, and the atmospheric pressure forces it back again. So soon as the return movement commences, a clutch contrivance engages the shaft and drives it. Engines of type 1A may be described as—

Engines using a gaseous explosive mixture at atmospheric pressure before explosion; the explosion acting on a piston free to move without connection with the crank shaft, the velocity being absorbed by the formation of a vacuum. The power is given to the shaft on the return stroke under the pressure of the atmosphere.

As has been stated in the historical sketch, the first to propose this kind of engine were Barsanti and Matteucci, 1857, but the difficulties were not sufficiently overcome until the invention of Otto and Langen, 1866.

Otto and Langen Engine.—This engine consists of a tall vertical cylinder surrounded by a water jacket; in it works a piston which carries a rack instead of a piston rod; the mouth of the cylinder is open to the atmosphere. Across the top of the cylinder is carried the fly-wheel shaft; it cannot be called the crank shaft because there is no crank. On the shaft there is a toothed wheel which engages the teeth of the rack; it runs freely on the shaft while the piston is on its upward stroke, but by an ingenious clutch arrangement it grips the shaft when the piston moves down. The shaft is therefore free to rotate in one direction and the piston is free to move up without restraint, but in moving down it gives the impulse. The shaft is carried on bearings bolted to the top of the cylinder, which forms a strong and convenient column for carrying the mechanism required to accomplish the cycle of the engine. At the lower end of the column is placed a slide valve which performs the treble duty of admitting, igniting, and discharging. It is driven from an intermediate shaft, intermittently, as determined by the governor of the engine. When working at full load, the movement of a small crank actuated from the shaft, lifts the rack and piston through some inches, taking in the charge through the slide valve, which then moves further and brings in the igniting flame. The explosion ensues and shoots up the piston with considerable velocity, the pressure rapidly falls by expansion and soon gets to atmosphere. The piston however has been moving freely and therefore has done no work; all the energy of the explosion, however, has been given to it. The piston has the energy of explosion in the form of velocity; it moves on, the pressure beneath it falling below atmosphere until all its energy of motion is absorbed in forming the vacuum. When this occurs it ceases its upward flight and returns, the outer atmosphere driving it back, and as the clutch has engaged the shaft, an impulse is given. The actual work is

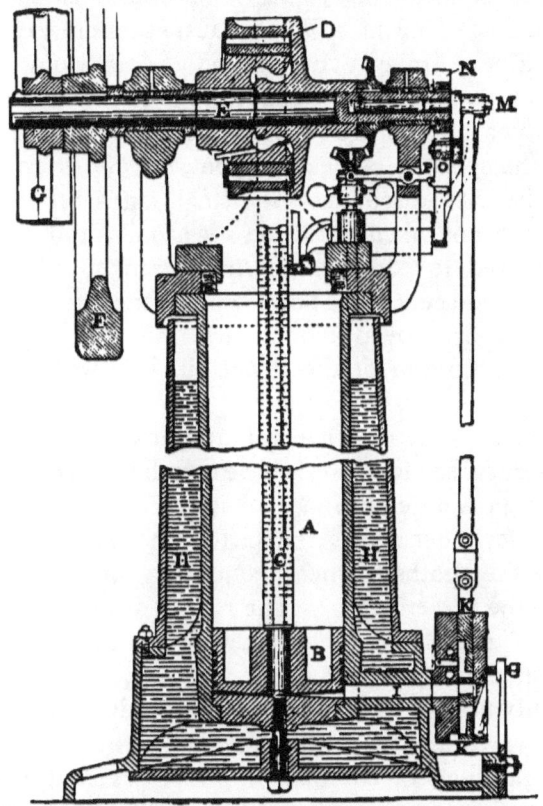

FIG. 32.—Otto and Langen Engine (vertical section).

therefore done by the atmosphere on the down stroke, the explosion being spent in obtaining energy in a form conveniently applicable. If no cooling of the hot gases occurred upon the down stroke the compression line would return to the point where the expansion line touched atmosphere; then the exhaust valve would open and the gases would be discharged at atmospheric pressure.

In that case the work done by the atmosphere and weight of piston on the downward stroke would exactly equal the energy of the explosion while falling by expansion to atmospheric pressure. But the cylinder does cool the gases while on the upward and downward stroke, so that the expansion line does not return upon itself; the amount of fall below the expansion line is gain and is added to the energy of the explosion just as the condenser adds to the efficiency of the expansion of steam. The exhaust gases are expelled by the piston and a new stroke is commenced. At full power the piston makes about 30 strokes per minute, the shaft rotating about 90 revolutions per minute. The governor of the engine is so arranged that when the speed becomes too great, a lever disengages a pawl from a ratchet and disconnects the small crank lifting the piston. The charge is not taken in till the speed falls, and then the pawl is again allowed to connect the small crank to the main shaft. The ignition slide gets its motion from the small crank shaft, so that it is arrested or moved along with the piston. The piston remains at the bottom of the stroke till it is wanted for another explosion.

Fig. 9, p. 21, shows the general arrangement of the engine, and fig 32 is a vertical section showing the clutch and section of the slide valve. Fig. 33 is an elevation, part in section.

A is the cylinder; B is the piston to which is attached the rack C; D the toothed wheel containing the clutch engaging the rack to the power shaft. The rack is strongly guided. E is the fly-wheel shaft on which is keyed the fly wheel F and the driving pulley G; H water jacket; I the port for inlet of the explosive mixture and discharge of the products of combustion; K the slide valve serving to admit, to ignite the charge and to discharge the products of combustion; it is actuated from the small shaft L by the pin M; the ratchet N and the pawl O connect the small shaft to the main shaft when requisite, as determined by the governor lever P.

This engine is the result of great care and labour on the part of the inventors; it is greatly superior in economy and efficiency to any preceding it, and its only fault is its excessive bulk and weight and the great noise made by it when in action. The whole of the energy of the explosion being expended in giving the piston velocity, just as in a cannon, the recoil is considerable. So serious is it that none but the very smallest engines can be placed upon upper floors without special strengthening. The author has seen an engine at work where the vibration produced was so great that props were put under the engine from floor to floor through four floors to get a solid resistance in the basement.

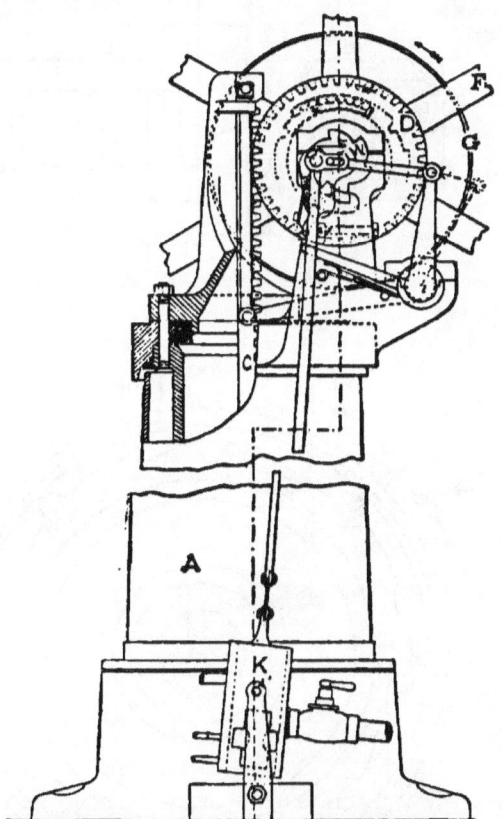

Fig. 33.—Otto and Langen Engine (elevation).

In other cases strong iron beams placed diagonally at the angle of a stone wall carried the engine; notwithstanding these precautions much vibration was caused. These difficulties did not seriously affect the sale of the engine for small powers, but they quite prevented it being made for powers above 3-horse. The clutch also is a matter of great difficulty, the whole power of the engine passes through it and it must act freely and instantaneously. The faintest back lash would allow the accumulation of so much velocity by the return that even a strong arrangement would be destroyed. For this reason the pawl and ratchet of Barsanti and Matteucci failed completely.

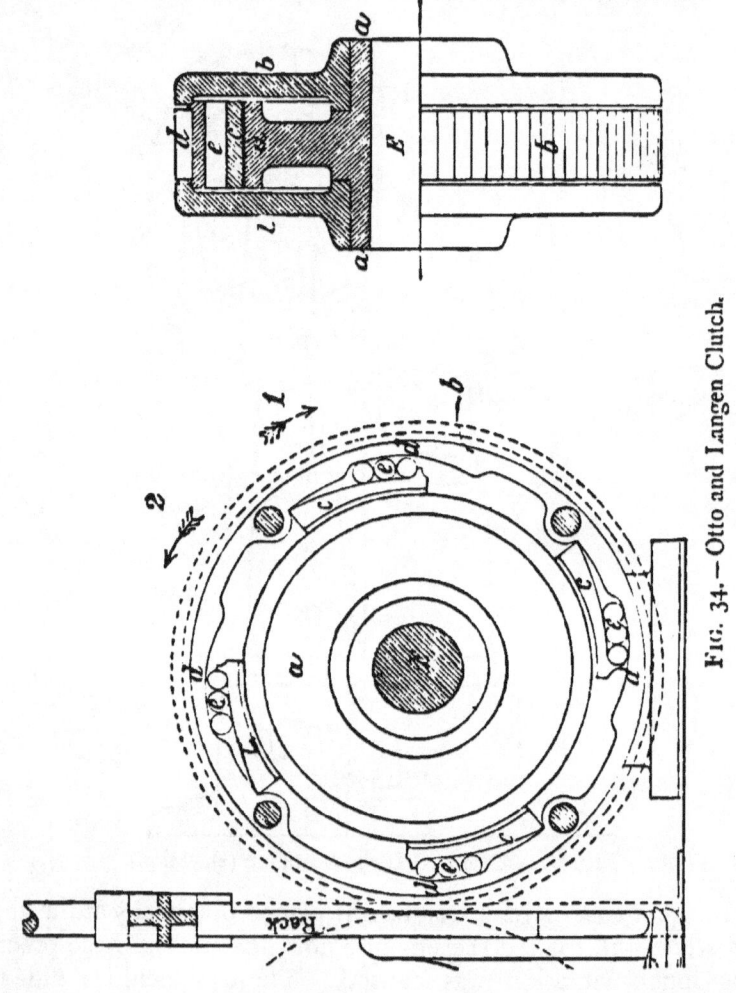

Fig. 34.—Otto and Langen Clutch.

Messrs. Otto and Langen's clutch is one of the main points of their invention and is excellent. It is shown in detail at fig. 34. The part *a* is keyed to the shaft; on it runs the part *b* carrying the teeth engaging the rack. So long as b moves in the direction of the arrow 1, or is stationary, a revolves freely with the shaft in the direction 2. The steel slips c, c, c, c are wedge-shaped on the back, so is the interior of the part b at the positions d, d, d, d. So long as the rack is stationary or ascending, the steel rollers e, e, e, e, run freely clear of the inclined surfaces; immediately the rack moves down at a rate greater than the movement at A, then the rollers are firmly wedged between the two inclined surfaces and the steel slips c, c, c, c grip the part a firmly and drive the shaft. When the bottom of the stroke is reached the wedges loose again and the piston is free.

Diagrams and Gas Consumption.—The author has made a set of experiments upon an engine of 2-horse power working with Oldham coal gas.

The cylinder is 12·5 inches diameter and the longest stroke observed was 40·5 inches. Working at the rate of 28 ignitions per minute, the indicated power was 2·9 horse, and the gas was consumed at the rate of 24·6 cubic feet per i.h.p. per hour. The brake power is 2 horse, so that the brake consumption is at the rate of 36 cubic feet per horse power per hour. This does not include the consumption of the side lights which is in all 12 cubic feet per hour.

Fig. 35 is a diagram from the engine when at full power.

The full line is that traced by the indicator, and the dotted line is the real line of pressures marred by the oscillation of the indicator pencil.

Professor Tresca tested a half-horse engine at the Paris Exhibition of 1867; it gave 0·456 brake horse, and consumed gas at the rate of 44 cubic feet per brake horse power per hour. This estimate did not include the side lights. The author's test gives a better result than that of M. Tresca, but this is due to the fact of the larger engine being used. It is probable that a 3-horse engine would give a consumption of about 30 cubic feet per brake horse power per hour.

The interest excited by the engine at the time of its first trial was naturally great, and many explanations were advanced of the cause of its superiority over the Lenoir and Hugon engines. Strangely enough the theory of the engine has been at best but imperfectly stated by previous writers; some indeed have fallen into grave error respecting its action. It is therefore essential that it should be somewhat fully considered here.

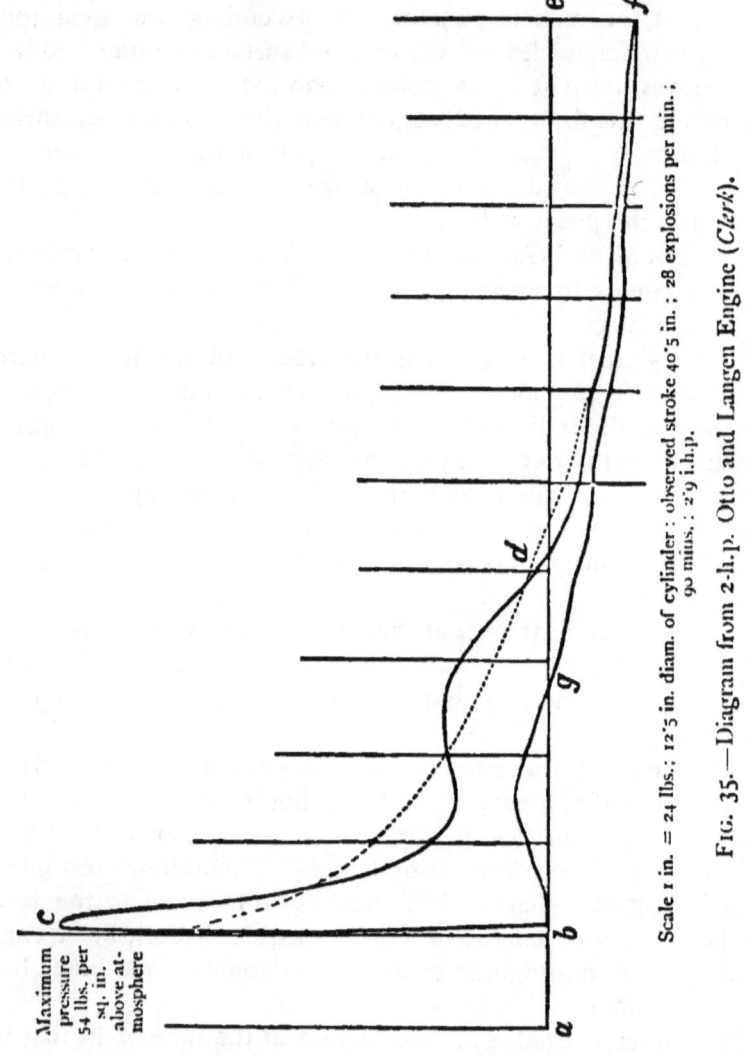

Fig. 35.—Diagram from 2-h.p. Otto and Langen Engine (*Clerk*).

Scale 1 in. = 24 lbs.; 12·5 in. diam. of cylinder: observed stroke 40·5 in.; 28 explosions per min.; 90 mins.; 2·9 i.h.p.

Maximum pressure 54 lbs. per sq. in. above atmosphere

The name by which it is most widely known is in itself misleading. Atmospheric gas engine at once suggests the Newcomen steam engine and further suggests the substitution of flame for steam, vacuum in both cases being supposed to be produced by condensation. In the steam engine the name truly describes the action: the piston is drawn up, the cylinder filled with steam at atmospheric pressure and the steam condensed by a water jet; then the atmosphere presses the piston down and gives the power.

In the gas engine cooling has little to do with the production of the vacuum; the vacuum would be produced and the engine would act efficiently without any cooling action of the cylinder whatever. The diagram fig. 35 proves this very clearly. While the piston is moving from the point a to b by the energy stored up in the fly wheel, the charge enters the cylinder; at b the piston pauses, and, the igniting flame being introduced, the charge explodes, the pressure rises to 54 lbs. per square inch above atmosphere. The appearance of the explosion curve does not indicate truly the rate of increase, because the piston is completely at rest till the pressure puts it in motion. The piston moves up impelled by the pressure of the explosion; as it moves the gases beneath it expand and therefore the pressure falls. At the point d the pressure is again level with that of the outside atmosphere; here the explosion ceases to impel the piston and, the pressure in the cylinder falling, the atmosphere presents a continually increasing resistance. But while the piston is passing from the point b to d, the pressure has been falling from 54 lbs. above atmosphere to atmosphere; the average pressure upon it through this distance is 12·6 lbs. per square inch; as the distance is 1·3 feet and the piston area is 122·7 inches, 2010 ft. pounds have been expended upon it. What becomes of this work? In an ordinary engine it would be communicated to the crank, and if no load were on, the crank would give it to the fly wheel. Here there is no crank and the piston is perfectly free, the piston alone contains the energy; its weight has been raised through 1·3 feet and the balance of the energy is stored in it as velocity of upward movement.

It must therefore continue to move up till its energy of motion is expended in compressing the atmosphere, in raising the piston, and in friction. If friction did not exist and the piston was indefinitely light, then the portion of the diagram bcd would be equal in area to the portion def, that is, the work expended by the explosion in giving the piston velocity would be equal to the work expended by the atmosphere in bringing it to rest again. Once at rest the vacuum produced allows the piston to be driven down again, this time to give up its energy to the motor shaft.

As the piston in this engine weighs 116 lbs. the work spent in raising it through 1·3 ft. is 116 × 1·3 = 150·8 ft. pounds; deduct this from the total work; and 2010 − 151 = 1859 ft. lbs. is the energy of motion of the piston.

The relation between energy, mass and velocity is

$$E = \frac{Mv^2}{2}.$$

E = energy in absolute units. One foot pound = 32 absolute units.

M = mass in pounds.

v = velocity in feet per second.

The velocity is therefore $v = \sqrt{\dfrac{2E}{M}}$

and

E = 1859 × 32 = 59488 absolute units.

M = 116 $v = \sqrt{\dfrac{2 \times 59488}{116}}$ = 32 nearly.

The velocity of the piston at the moment when the explosion pressure has been expended and the internal and external pressures exactly balance is 32 ft. per second or 1560 ft. per minute; at no point of the stroke in any ordinary engine, steam or gas, is such a high piston speed possible. This explains the recoil of the engine. But this is not the average. The piston has attained 32 feet per second after moving through 1·3 feet; the time taken to move that distance is $t = \sqrt{\dfrac{2s}{v}}$ when t = time in seconds.

s = space passed through.
v = velocity.

and s = 1·3 feet v = 32 feet.

$$t = \sqrt{\frac{2 \times 1·3}{32}} = 0·28 \text{ second.}$$

The piston has taken 0·28 second to move through the 1·3 feet; its average velocity during the action of the explosion is therefore 4·64 feet per second or 278 feet per minute. This, although high, is not greatly in excess of that used in the Lenoir and Hugon engines. It is less, indeed, than the average piston speed now used in modern compression engines, 300 to 400 feet per minute being common. If no cooling by the cylinder occurred, the line cdf would be adiabatic, and the return line fg would coincide with the expansion line df: the portion of the vacuum diagram def is due solely to the energy of the explosion, the part dfg is due to the cooling of the gases. If cooling did not act at all, the area bcd would be greater, and therefore def, which is its equivalent, would also be greater, that is, the vacuum produced would be greater if no cooling whatever existed.

The theory of its action generally held at the time of M. Tresca's experiments seems to have been as follows :

The work of the explosion consists simply in pushing up the piston and filling the space behind it with flame, which flame is cooled by contact with the cylinder, and a vacuum results. The flame is considered as analogous to steam, and the cooling as similar to condensation as in the Newcomen engine. The inventors of the engine seem to share this erroneous idea; certainly M. Tresca did, as in his report upon the engine he says : 'There is, therefore, between the older machines and the new one this difference of principle, that the pressure in the cylinder can never descend below the atmospheric during the upward stroke. The negative force of the atmospheric pressure, useless as it was, becomes utilisable. . . .' He clearly considered that the pressure during the upward stroke was expended only in lifting the piston through a certain height, and as soon as it fell to atmosphere the piston stopped and the cooling caused a vacuum, the work being done by the falling of the piston and the pressure of the exterior air. If the cooling really caused the vacuum, the diagram would be quite different; instead of the pressure touching atmosphere at the point d, it would not touch till the point e at the end of the stroke. The pressure would then abruptly fall to f, and the piston would return. M. Tresca observed that the pressure did fall below atmosphere before the end of the stroke, but he considered it as a defect. 'In reality the piston rises in virtue of the swiftness acquired to beyond the position at which there would be equilibrium between the interior and exterior pressure. But that one small loss of power is amply compensated for by the atmospheric power of the downward stroke.'

The very principle of the engine really depends upon this fall of pressure which was considered by M. Tresca a defect; it is the only way to store up the power of the explosion so that it may be available on the downward stroke. If the pressure did not fall, the piston would require to be projected 10·5 feet into the air to absorb the energy of the explosion in its mass alone; by the fall of pressure it is absorbed with the smaller movement of 1·8 feet. The only part of the diagram due to cooling is the part dfg, not more than one-fifth of the total area representing work done by the engine.

The superior economy, it is evident, cannot be altogether due to greater piston velocity; the piston velocity, although considerable, is not superior enough to that of Lenoir and Hugon to account for all the difference. There must exist other points of dissimilarity. In the Lenoir type of engine the strokes were numerous and the gas consumed per stroke on the whole smaller than in Otto and Langen engines of equal power: the latter used few strokes but large cylinders; proportionally the cooling surface exposed was thus diminished. Then the piston is at rest until the explosion puts it in motion; the pressure gets time to rise to its maximum before the piston moves and expands the space. Maximum pressure is attained at constant volume as required by theory; at the same time the piston and cylinder remain cool because of the infrequency of the strokes. The entering charge is therefore but slightly heated before explosion, and the explosion gives a better pressure for a smaller elevation of temperature.

The most potent cause of improvement, however, is great expansion: the large cylinders allow an expansion of 10 times the volume existing before explosion, and so gain, first by expanding to atmosphere, and second by the cooling which follows the further expansion. A comparison of the actual diagram with the theoretical reveals some interesting peculiarities which seem hitherto to have escaped observation. The maximum pressure on the diagram, which is above the true pressure, fig. 35, is 54 lbs. above atmosphere, corresponding to a temperature of 1355° absolute.

The mixture exploded contains 1 volume gas and 7 volumes air (Oldham gas); if all the heat present had been evolved by the explosion the pressure should have been 168 lbs. above atmosphere. At the maximum pressure only 32 per cent. of the heat has been evolved, leaving 68 per cent. to be evolved during the expansion. The line cd is very much above the adiabatic, so much so that the curve cd is nearly isothermal, the temperature at d only becoming 1305° absolute instead of 733°, which it should be if adiabatic. The heated gases are therefore gaining heat from c to d, and as the only source is combustion, it follows that the combination is not nearly complete at the maximum pressure. The 68 per cent.

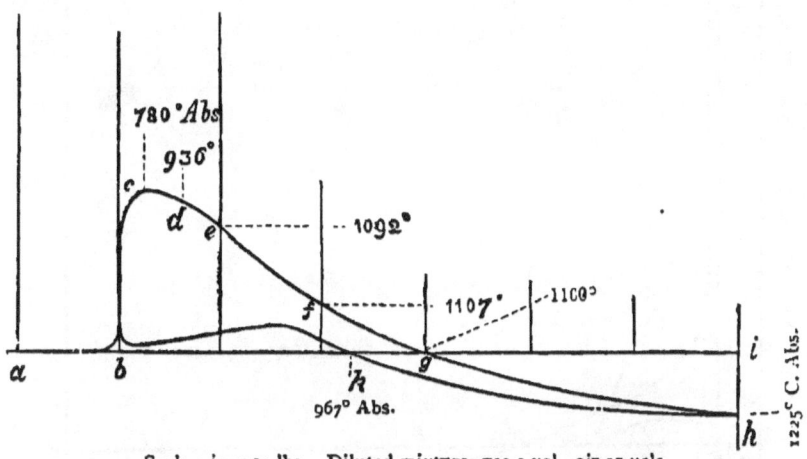

Scale 1 in. = 24 lbs. Diluted mixture, gas 1 vol., air 12 vols.

FIG. 36.—Diagram from 2 h. p. Otto and Langen Engine (Clerk).

of the total heat which has not appeared at the maximum pressure is appearing during the expansion. The combustion seems to be nearly complete at the point d as the line de behaves as if cooling; if adiabatic, the temperature at f should be 961°—it is 870°. During compression to atmosphere again, the temperature remains constant at 870°; the cooling power of the cylinder is equal only to preventing increase which would otherwise occur.

This effect is more evident with a more dilute mixture. Fig. 36 is a diagram taken by the author from the same engine, but using a mixture containing 1 volume of gas and 12 volumes of air. Here the maximum pressure is only 17 lbs. per square inch above atmosphere. With complete evolution of heat it should be 103 lbs.; the maximum pressure in this case only accounts for 24 per cent. of the heat known to be present, leaving 76 per cent. to be evolved during expansion. The diagram affords the most ample proof that the

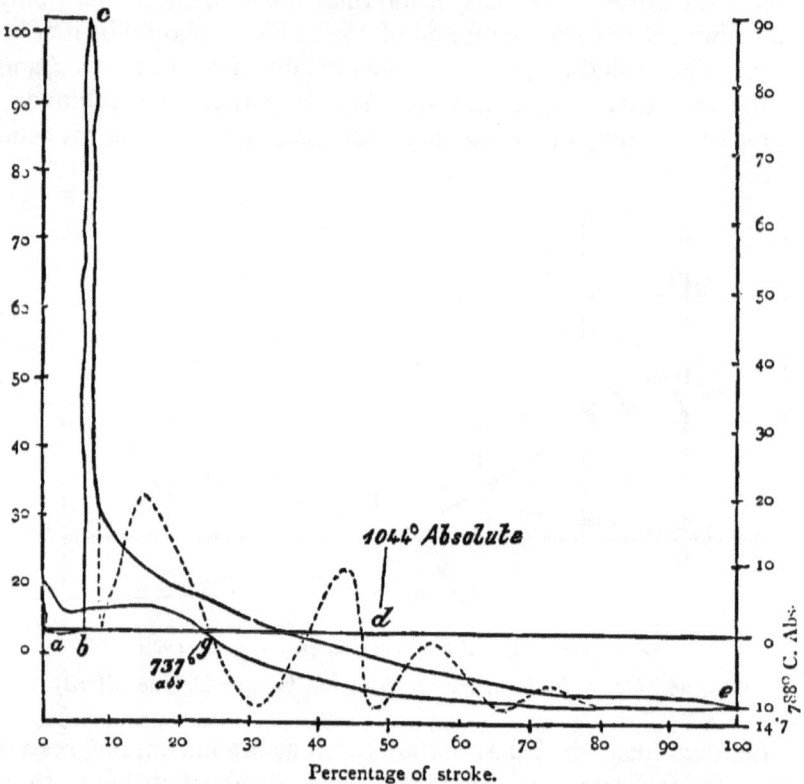

FIG. 37.—Otto and Langen Engine. Free Piston.

combustion is proceeding, the falling line shows a steady increase of temperature to the very end of the stroke. The temperatures are marked upon the diagram at the successive points; taking the temperature at the point b as 290° absolute, the points c, d, e, f, g, h are respectively 780°, 936°, 1092°, 1107°, 1160°, and 1225°, showing a steady increase throughout the whole expansion line, right to the end of the stroke. The consumption of gas per indicated HP rises very much in consequence, amounting to about 37 cubic feet per IHP hour. The power at the same time falls, so that the 30 explosions per minute are required to keep the engine going without load at 53 revolutions per minute. The cooling during compression is so slow that the temperature falls only from 1225° to 967°, from the point h to k. All the published diagrams examined by the author show this peculiar effect. Fig. 37 is a diagram published by Mr. F. W. Crossley. Taking 80 lbs. as the maximum pressure, which seems somewhat higher than is warranted by the diagram, the oscillation of the indicator has been so excessive, the corresponding temperature is 1873° absolute : the expansion line cd if adiabatic would give at the point d a temperature of 1090°, the actual temperature is 1044°. Within the limits of error they may be considered the same ; there is therefore combustion going on from c to d also. At the point e the temperature is 788° ; if adiabatic it should be 667°. It is quite evident that the whole of this expansion curve is above the adiabatic ; in the earlier part of the diagram the oscillation causes uncertainty, but in the latter part the measurement is true enough.

The compression line eg is almost isothermal, 788° at e, cooling to 737° at g.

Fig. 38 is a diagram by Releaux taken from Schöttler.

It is manifestly wrong, as the vacuum part is much too small and the maximum temperature is higher than has ever been obtained by any explosion of gas and air, but if taken as relatively correct the expansion line is much above the adiabatic.

From his study upon the explosion of gas and air mixtures in closed vessels, the reader will be prepared to find that only a portion of the total heat present is evolved by explosion in any gas engine. That is, the explosion maximum pressure never accounts for the whole heat present as inflammable gas ; a portion is in some manner suppressed and is not evolved till long after the moment of complete explosion. Combustion is not completed till considerably after the completion of explosion.

He will be unprepared, however, for such diagrams as figs. 35 and 36, where the maximum pressure represents only 0·32 and 0·24 of the heat present, and 0·68 and 0·76 are evolved during the forward stroke while the pressure is falling. The explanation is simple. The case is quite different from that of the closed vessel or where the piston is connected to a crank. As soon as the pressure of the explosion becomes great enough, the piston at once moves out and prevents further increase of pressure. The slower the rate at which the mixture inflames, the greater will be the apparent suppression of heat; thus the mixture 1 volume gas 7 air takes, in the closed vessel, 0·06 second to complete the explosion, but before this

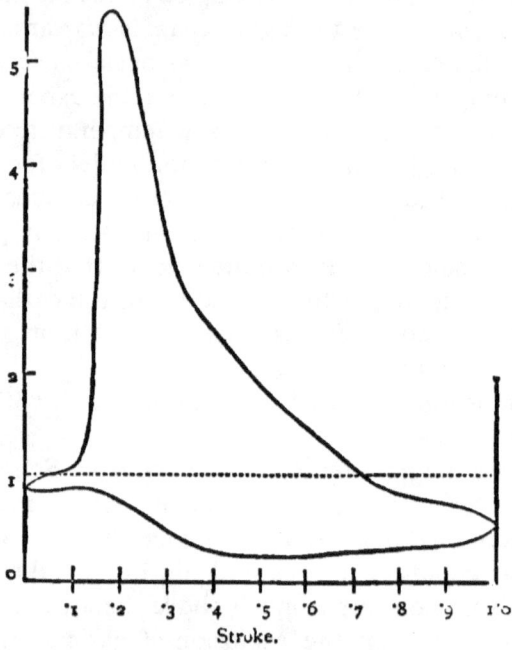

FIG. 38.—Otto and Langen.

time has elapsed the piston is in rapid motion reducing the pressure before complete explosion. To get the maximum pressure it would be necessary to prevent it from moving till the explosion was complete. The weaker mixture takes 0·25 second to complete the explosion, and so in diagram, fig. 36, the temperature actually rises throughout the whole stroke.

A heavier piston would be longer in starting under the pressure of the explosion and would so allow a high pressure to be attained

Gas Engines of Different Types in Practice

with a given mixture. This explains Mr. Crossley's remark in reading his paper, that heavy pistons gave a more economical result than light ones.

Gilles Engine.—The great success of the Otto and Langen engine occasioned many attempts to improve upon it. Its merits and its faults were equally evident. The recoil of the engine at every stroke was exceedingly troublesome, and the noise of the rack and clutch could be heard at a long distance. Gilles of Cologne invented an engine intended to retain the economy while reducing the noise. In it there are two pistons, one free, the other connected to the crank in the usual way. The free piston being at the bottom of its stroke and close to the crank piston, the latter moves a portion of its stroke, taking in the explosive charge ; at a suitable position ignition occurs and the free piston is driven in one direction while the other completes its outstroke. A vacuum is produced between the pistons, and the free piston rod being gripped by a clutch is kept in its extreme position till the main piston returns under the pressure of the atmosphere. The clutch is then released and the free piston falls, expelling the exhaust gases. Fig. 39 is a

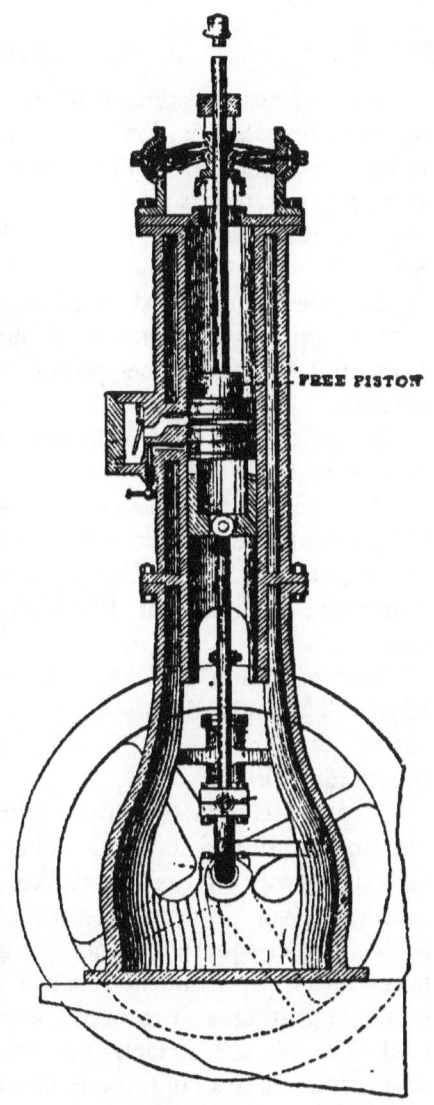

Fig. 39.—Gilles Engine. Free Piston.

section of the engine. It is unnecessary to give further description, as the engine was not so economical or so simple as the earlier one.

Type II.

In engines of this kind compression is used previous to ignition, but the ignition is so arranged that the pressure in the motor cylinder does not become greater than that in the compressing pump. The power is generated by increasing volume at a constant pressure. Engines of Type II. are therefore :

Engines using a mixture of inflammable gas and air compressed before ignition and ignited in such a manner that the pressure does not increase, the power being generated by increasing volume.

These engines are truly slow combustion engines ; in them there is no explosion.

The most successful engine of the kind is an American invention; although proposed in 1860 by the late Sir William Siemens, it was never put into practicable workable shape till 1873, when the American, Brayton of Philadelphia, produced his well-known machine.

Messrs. Simon of Nottingham introduced it into this country in 1878. They added one thing only of doubtful utility—that is, the use of steam raised in the water jacket as auxiliary to the flame in the motor cylinder.

Brayton Engine.—In this engine there are two cylinders, compressing pump and motor. The charge of gas and air is drawn into the pump on the out-stroke and compressed on the return into a receiver ; the pressure usual in the receiver varies from 60 to 80 lbs. per square inch above atmosphere. The motor cylinder takes its supply from the receiver but the mixture is ignited as it enters, a grating arrangement preventing the flame from passing back; the mixture, in fact, does not enter the motor cylinder at all; what enters it, is a continuous flame. At a certain point the supply of flame is cut off and the piston, moving on to the end of its stroke, expands the volume of hot gases to nearly atmospheric pressure before discharge.

Gas Engines of Different Types in Practice 153

Fig. 40 is an external view of the engine. Figs. 41 and 42 are sections of the motor and pump cylinders. The action is as follows :—The engine is single acting, receiving one impulse for every revolution ; like all gas engines it depends upon the energy stored up in the fly wheel to carry it through those parts of its cycle

FIG. 40.—Brayton Petroleum Engine.

where the work is negative. The two cylinders are inverted and are attached to a beam rocking beneath them, by connecting rods. The beam is prolonged and connected to the crank above it by a rod ; both cylinders are single-acting and the pistons are of the trunk kind. Both pump and motor cylinders are of the same diameter, but the pump is only half the stroke of the motor. The valves are actuated from a shaft running at the same rate as the main shaft and driven from it by bevel wheels. There are four valves, all of the conical seated kind—two upon the motor, admission and discharge, two upon the pump cylinder, admission and discharge. The admission and discharge valves upon the motor are actuated from the auxiliary shaft by levers and cams, so is the pump inlet. The pump discharge valve is automatic, rising at the proper time by the pressure of compression. During the down-stroke the pump takes in the charge of gas and air, forcing it on the up-stroke into the receiver. From the receiver it is led to the power cylinder, passing by the inlet valve through a pair of perforated brass plates with wire gauze placed between them. Through this diaphragm a small stream of mixture is constantly passing into the motor cylinder; before the engine is started, a plug is withdrawn and the current lighted; a constant flame is therefore burning under the diaphragm. The mixture enters the cylinder through this flame, lighting as it enters; at all times during the exhaust part of the stroke, as well as the admission, the stream of entering mixture, from the receiver, keeps up a small constant flame which is augmented at the beginning of the stroke, so as to fill the cylinder entirely, when the admission valve is opened. When the admission valve is closed, the bye-pass keeps the flame fed with sufficient mixture to keep it alight. The pressure in the cylinder thus never exceeds that in the reservoir and the mixture burns quietly without spreading back.

Figs. 40, 41 and 42.—A is the motor cylinder; B the pump; the beam and connections require no lettering; c is the pump inlet valve (the pump discharge, which is an ordinary lift valve, is not seen in fig. 42, but is lettered D in fig. 40); E the motor inlet; F the igniting plug which is withdrawn when the flame is to be lit before starting the engine (see fig. 82); G is the grating in section (see fig. 41); H the exhaust valve; the levers and cams are sufficiently indicated on the drawing; the small pipe and stop-cock I (fig. 40) communicates at all times with the reservoir and supplies the constant flame with mixture. The engine worked well and smoothly; the action of the flame in the cylinder could not be distinguished from that of steam, it was as much within control and produced diagrams quite similar to steam. The flame grating was the weak point; it stood exceedingly well for a time, but if by any accident the gauze was pierced in cleaning, the flame went back into the reservoir and exploded all the mixture—the engine, of course, pulled up as the constant flame having no supply

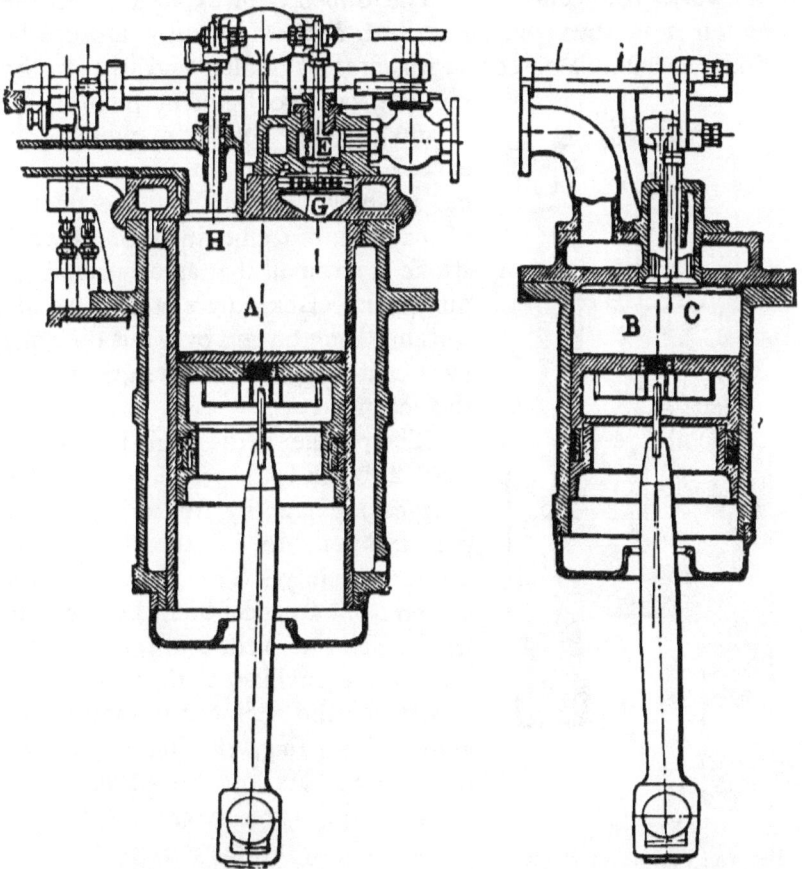

FIG. 41.—Brayton Engine. Section of Motor Cylinder.

FIG. 42.—Brayton Engine. Section of Pump Cylinder.

was extinguished. This accident became so troublesome that Mr. Brayton discontinued the use of gas and converted his engine into a petroleum engine. The light petroleum was pumped upon the grating into a groove, filled with felt, the compressing pump then charged the reservoir with air alone. The air in passing through the grating carried with it the petroleum, partly in vapour, part in spray; the constant flame was fed by a small stream of air. The arrangements were, in fact, precisely similar to the gas engine, except in the addition of the small pump and the slight alteration in the valve arrangements. The difficulty of explosion into the reservoir was thus overcome, but a new difficulty arose—the cylinder accumulates soot with great rapidity and the piston requires far too frequent removal for cleaning. The petroleum pump is an exceedingly clever little contrivance; fig. 43 shows its details. The amount of petroleum to be injected at each stroke is so small that an ordinary force pump with clack valves would be uncertain. Brayton gets over this difficulty by substituting a slide valve driven from the eccentric.

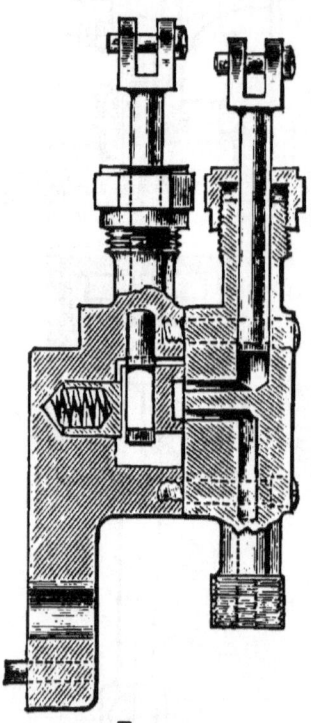

FIG. 43.
Brayton Petroleum Pump.

The plunger of the pump is no larger than a black-lead pencil, yet it discharges any quantity, from a single drop per stroke up to full throw, with unerring certainty. The plunger also is driven from an eccentric. Both eccentrics are in one piece and rotate on the end of the auxiliary shaft, driven by a pawl when the engine is in motion; to allow of starting, the pump can be moved by a hand-crank independently. To start, the air reservoir is filled, if not already full, by turning the engine round by hand; the plug F is then withdrawn and a little petroleum thrown upon the diaphragm by a few turns of the pump. The cock I on the small pipe is then opened and a stream of air flowing from the reservoir vaporises the petroleum; it is lit at G, and the flame having enough air for combustion retreats to the grating and remains burning within the cylinder. The plug is then inserted, the starting cock opened, and the engine starts.

The flame remains alight during the whole time the petroleum continues to be supplied.

The valves act well and the motor cylinder does not suffer from the action of the flame so long as it is kept reasonably clean. If the soot, however, is allowed to accumulate, it speedily cuts up.

Diagrams and Gas or Petroleum Consumption.—Prof. Thurston of the Stevens Institute of Technology tested a Brayton gas engine in New York in the year 1873.

The following extracts are from his report :

'The operation of the engine is precisely similar in the action of the engine proper and in the distribution of pressure in its cylinder, to that of the steam engine. The action of the impelling fluid is not explosive as it is in every other form of gas engine of which I have knowledge.

'Upon the opening of the induction valve, the mixed gases enter, steadily burning as they flow into the cylinder, and the pressure from the commencement of the stroke to the point of cut off, as is shown by the indicator diagrams, is as uniform as that observed in any steam engine cylinder. The maximum pressure exerted during my experimental trial, and while the engine was driving somewhat more than its full rated power, was about 75 lbs. per square inch at the beginning of the stroke, gradually diminishing to 66 lbs. per square inch at the point of cut-off, where the speed of the piston was nearly at a maximum, and then declining in accordance with the law governing the expansion of gases.

'Complete combustion is insured by thorough mixture. This is accomplished by taking the illuminating gas and air, in proper proportion, into the compressing pump together, and the mixture here made becomes more intimate in the reservoir, and in its progress towards the point at which it does its work. The constantly burning jet already described insures prompt ignition on entering the cylinder.

'. . . the engine rated at 5 HP developed, as a maximum, rather more than its rated power. Its mean power during the test, as determined by the dynamometer, was 3·986 HP, the indicator showing at that time 8·62 HP developed in the cylinder. The amount of gas consumed averaged 32·06 cubic feet per indicated HP per hour.

'The excess of indicated over dynamometric HP is to be attributed to the work of driving the compressing pump and to the friction of the machine.

'The greater portion of this appears both in debit and credit side of the account, since, although expended in the compressing pump, it is restored again in the driving cylinder.'

The consumption of 32·06 cubic feet per horse hour is incorrect; it is obviously unfair to include the pump diagram in the gross power. The author has tested an engine of similar construction and dimensions; he finds the friction of the mechanism

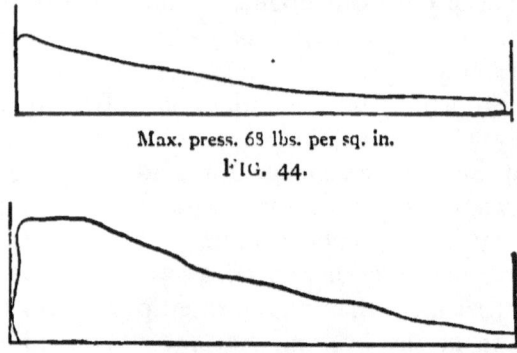

Max. press. 68 lbs. per sq. in.
FIG. 44.

FIG. 45.—Diagrams from Brayton's Gas Engine.

to be about 1-horse; adding this number to the dynamometric power of Prof. Thurston, the legitimate indicated power may be taken as 5 HP, the consumption is therefore $\frac{8\cdot62 \times 32\cdot06}{5\cdot0} = 55\cdot2$; and the gas per brake HP per hour is $\frac{8.62 \times 32\cdot06}{3\cdot986} = 69\cdot3$. These numbers, although showing improvement upon the Lenoir and Hugon, prove that the engine was much inferior in economy to the Otto and Langen engines.

Mr. H. McMutrie, Consulting Engineer at Boston, took diagrams from an engine of similar dimensions which confirm these results. Fig. 44 is the diagram taken with full load, fig. 45 the diagram from the motor with no load on, the power being just sufficient to overcome friction and pump losses.

Full Load Diagram.

Area of piston	50·26 sq. ins.
Speed of piston	180 ft. per min.
Mean pressure	33 lbs. per sq. in.
Pressure in reservoir	75·4 lbs. per sq. in.
Initial pressure in cylinder	68 lbs. per sq. in.
Gross power developed	9 HP.

No Load Diagram.

Speed of piston	180 ft. per min.
Mean pressure	18 lbs. per sq. in.
Friction and other resistance	4·87 HP.
Net available power	$9 - 4·87 = 4·13$

This power agrees closely with the actual determination by dynamometer.

The author has made a careful trial of a Brayton petroleum engine rated at 5-horse. The engine was made by the 'New York and New Jersey Ready Motor Company;' it was sent to Glasgow and the following test was made at the Crown Ironworks on the 21st and 22nd February, 1878. The motor cylinder is 8 inches in diameter and the stroke 12 inches; the pump cylinder is also 8 inches diameter but the stroke is 6 inches.

Diagrams were taken from both pump and motor by a well-made Richards' indicator. At the same time the dynamometer was applied to the fly wheel fully loading the engine, readings were taken at regular intervals. The revolutions were recorded by a counter. The petroleum used was measured in a graduated glass vessel.

The results are as follows:

Test of Brayton Petroleum Engine. (*Clerk.*)

Petroleum consumed during one hour	1·378 gallons.
Mean speed of engine	201 revs. per min.
Mean dynamometer reading	4·26 HP.
Mean pressure, power cylinder	31 lbs. per sq. in.
Mean pressure, air pump	27·6 lbs. per sq. in.
Piston speed, motor	201 ft. per min.
Piston speed, pump	100·5 ft. per min.
Power indicated in motor	9·49 HP.
Power indicated in pump	4·10 HP.
Available indicated power	5·39

The power by the dynamometer is 4·26-horse; therefore the mechanical friction of the engine is $5·39 - 4·26 = 1·13$ horse.

Consumption of petroleum . . . 0·255 galls. per IHP per hr.
Consumption of petroleum . . . 0·323 galls. per actual HP per hr.

Figs. 46 and 47 are diagrams from the motor and pump, which are fair samples of those taken. It will be observed that considerable throttling occurs in entering the motor cylinder; the pump pressure is higher than the reservoir pressure, and the motor pressure is lower, so that a double loss has been incurred. The principle of the engine is so good that the author anticipated better results. Great improvement could be obtained by reproportioning

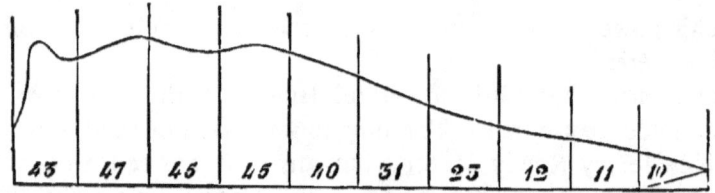

Mean pressure 30·2 lbs. per sq. in. 8 ins. dia. cylinder. Stroke 12 ins. 200 revs. per min.
FIG. 46.—Brayton Petroleum Engine. Motor Cylinder.

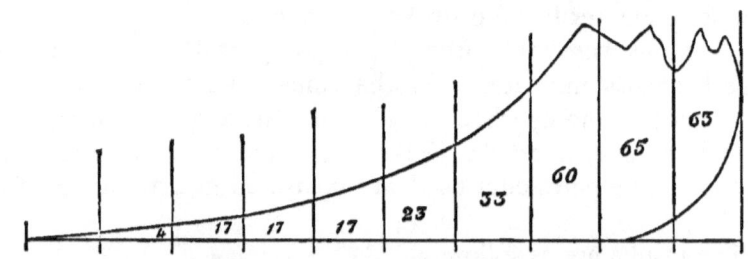

Mean pressure 27·6 lbs. per sq. in. 8 ins. dia. cylinder. Stroke 6 ins. 200 revs. per min.
FIG. 47.—Brayton Petroleum Engine. Pump Cylinder.

the valves and air passages; they are in this engine much too small and cause needless resistance and loss. The maximum pressure in the motor cylinder is 48 lbs. per square inch, which remains steadily till the inlet valve shuts at four-tenths of the stroke: the pressure then slowly falls as the gases expand, the exhaust valve opening at about ten pounds per square inch above atmosphere.

The average available pressure upon this diagram is 30·2 lbs.

per square inch. The air pump shows a maximum pressure of 65 lbs. per square inch, the reservoir pressure being 60 lbs. The average resistance is 27·6 lbs. per square inch; as the pump is half the stroke of the motor and equal to it in area, the pressure to be deducted is $\frac{27·6}{2} = 13·8$ and $30·2 - 13·8 = 16·4$. The actual available pressure actuating the engine is therefore only 16·4 lbs. per square inch. The effect of the clearance in the pump cylinder is noticeable upon the diagram; the air inlet valve does not open till one-tenth of the down stroke is completed.

The theoretic efficiency of this type, with a maximum temperature of 1600° C., compression of 60 lbs. per square inch above atmosphere, and motor cylinder of twice the pump volume, is 0·30; the efficiency of the gas in the mixture commonly used, 1 volume gas 7 volumes of air, is 0·40 (p. 113); so that if the conditions of loss by cooling are no worse than in the author's explosion experiments, and the diagram appeared perfect, the actual indicated efficiency would be $0·30 \times 0·40 = 0·12$. That is, the engine should convert 12 per cent. of the heat it gets as gas or petroleum into indicated work. But the diagram is imperfect in many ways. Using the mixture it does, the diagram should show a maximum temperature of 1600° C. at least; in reality the highest temperature is only 840° C. The flame is entering the cylinder at an actual temperature of 1600° C. during the whole period of admission, but the convection has so greatly increased by the mixing effect of the entering current that greatly increased cooling results; accordingly, when the gases are fully admitted and the inlet valve is closed, the gases have only a temperature of 840° C. instead of 1600° C. After admission ceases, the expansion line from 45 lbs. to 10 lbs. pressure is far above the adiabatic, indeed it is isothermal, the combustion is proceeding and the small igniting flame also is helping to sustain the temperature.

It is therefore quite evident that the loss of heat is much greater than that occurring during explosion in equal time. The correction of the theoretic efficiency indicated by the author's closed vessel experiments is insufficient, 0·12 is much above the actual efficiency. Taking the heating value of the American coal gas used in Prof. Thurston's experiments as 10,900 heat units per unit weight of gas burned, and one pound of it as measuring 30 cubic feet, then as the engine used 55 cubic feet per IHP per hour, its efficiency is 0·071; that is, it converts 7·1 per cent. of the heat given to it into work.

This is a poor result for a cycle having so high a theoretic efficiency, and in the author's experiments with petroleum it is even worse.

The sp. gravity of the petroleum was 0·85, therefore the weight of one gallon is 8·5 lbs. As 0·255 gallons are burned per indicated horse power per hour, this amounts to $8·5 \times 0·255 = 2·16$ lbs. of liquid fuel per IHP per hour. One pound gives out 11,000 heat units, and for one horse power for one hour 1424 units are required; the actual indicated efficiency is therefore

$$\frac{1424}{2·16 \times 11000} = \frac{1424}{23760} = 0·06 \text{ nearly}$$

; that is, 6 per cent. of the whole heat given to the engine is accounted for by the power developed in the motor cylinder.

If there were no losses of heat to the cylinder, or losses by throttling during the inlet and transfer of the air from the pump to the motor or loss of heat from the reservoir to the atmosphere, then the efficiency of this type of engine would be 30 per cent. These losses in practice reduce it to 6 per cent. The cycle is a good one, and under other circumstances is capable of better things, but it is quite unsuitable for a cold cylinder engine. Cooling and undue resistance are the main causes of the great deficit.

The gases entering the cylinder as flame, in passing through the inlet chamber expose a large surface to the action of the water jacket; the entering currents also impinge against the piston, causing more rapid circulation than ordinary convection. Both causes intensify the cooling action of the cylinder walls. In the engine tested by the author the communicating pipes and the motor admission valve were much too small; a considerable loss of pressure resulted; although the reservoir pressure was 60 lbs., that in the cylinder never exceeded 48 lbs. above atmosphere, showing a loss of 12 lbs. per square inch from undue resistance. To enable this engine to realise the advantages of its theory considerable modifications in its arrangements are required. Notwithstanding all difficulties it has done much useful work, not the least notable being the assistance it rendered to Prof. Draper during his investigation on the existence of non-metallic bodies in the sun's atmosphere. He used a Brayton petroleum engine for driving his dynamo machine, and he stated in his paper that its ease of starting and almost absolute steadiness in driving were of the greatest service to him. In steadiness he states that 'it acted like an instrument of precision.'

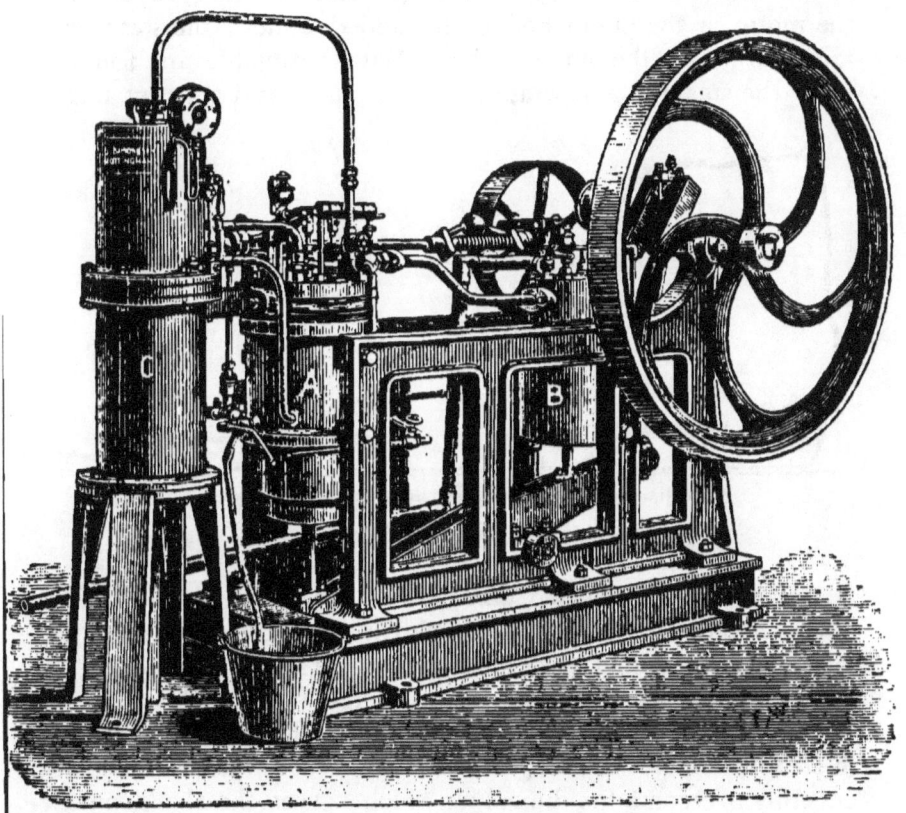

FIG. 48.—Simon Engine.

Simon Engine.—Messrs. Simon, of Nottingham, introduced the Brayton engine to England in a slightly altered form as a gas engine. In addition to the ordinary arrangements of the engine they attempted to gain increased economy, by causing the waste heat passing into the water jacket, and the heat of the exhaust gases, to be utilised in raising steam. They would undoubtedly have increased the economy of the engine in this manner had they not turned the steam so raised into the motor cylinder along with the flame. The cooling of the flame which was serious enough in the original was thus made worse, and but slight gain could result, the loss by cooling being slightly exceeded by the increase of volume due to the steam. Fig. 48 is an external view of the engine as exhibited at the Paris Exhibition of 1878. A is the motor, B the pump, and C the added boiler; the steam was raised in it and the water jacket. With a suitable arrangement using the steam in a separate cylinder, doubtless 6 per cent. might

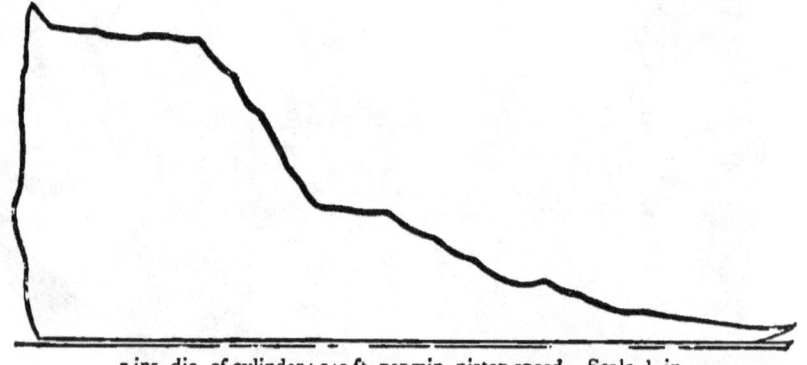

7 ins. dia. of cylinder; 240 ft. per min. piston speed. Scale $\frac{1}{14}$ in.

FIG. 49.—Diagram from Simon Engine.

be added to the indicated efficiency of the engine, but it is very questionable if the increased complexity does not entirely destroy any advantage gained; it certainly does so in small engines. When very large engines come to be constructed the complexity would not be so great and it would be well worth while to use waste heat in steam raising. The engine, although instructive, did not successfully overcome the difficulties which caused the abandonment of the Brayton as a gas engine. Fig. 49 is a diagram from the engine which forcibly illustrates the effect of the cooling.

Type III.

Engines of this kind resemble those just discussed, in the use of compression previous to ignition, but differ from them in igniting at constant volume instead of constant pressure; that is, the whole volume of mixture used for one stroke is ignited in a mass instead of in successive portions.

The whole body of mixture to be used is introduced before any portion of it is ignited; in the previous type the mixture is ignited as it enters the cylinder, no mixture being allowed to enter except as flame. In Type III. the ignition occurs while the volume is constant; the pressure therefore rises; it is an explosion engine in fact, like the first type, but with a more intense explosion due to the use of mixture at a pressure exceeding atmosphere.

The most obvious means of applying the method is that suggested by the Lenoir engine. The addition of a pump taking mixture at atmospheric pressure, compressing it into a reservoir from which it passes to the motor cylinder at the increased pressure, seems a simple matter. The igniting arrangements would act as in the original. As the gases are under pressure, the piston would take its charge into the cylinder in a smaller proportion of the forward stroke, and so more of the motor stroke would be available for useful effect. The diagram such an engine should produce is seen at fig. 15, p. 50; the shaded part is the available portion, the other part is the pump diagram. The theoretic efficiency of such an engine is as good as the type can give. The patent list shows that it was the first proposed after Lenoir. Many such engines have been attempted and have given very good results economically, but the difficulties of detail are considerable, the greatest being the necessity for the intermediate reservoir. Million's patent 1861 proposes to do this, the present author also constructed one of this kind in 1878, and later one was made by Mr. Atkinson. The difficulties, however, are too great to allow the success of small motors on the plan.

Mr. Otto, the first to succeed with the free piston engine, was also the first to succeed in adapting compression in a reliable form.

In the third type are included all engines having the following characteristics, however widely the mechanical cycle may vary :

Engines using a gaseous explosive mixture, compressed before ignition, and ignited in a body, so that the pressure increases while the volume remains constant. The power is obtained by expansion after the increase of pressure.

Otto Engine.—In this gas engine, the first to combine the compression principle with a simple and thoroughly efficient working cycle, the difficulties of compression are overcome in a strikingly original manner. To the engineer accustomed to the steam engine, the main idea seems a bold and indeed a retrograde step. The early gas engines were moulded more upon the steam engine model and were to some extent double acting. The Lenoir and Hugon both received two impulses for every revolution, the Brayton was single acting, and the Otto is only half single acting. The steam engine in its advance passed from single to double acting, and then to four and even more impulses per revolution. The gas engine in its progress has in this respect moved backwards, beginning with double action and then going back. The gain of this arrangement, however, has completely justified the retrogression.

In external appearance the engine closely resembles a modern high pressure steam engine, the working parts of which are of somewhat excessive strength ; its motor and only cylinder is horizontal and open ended ; in it works a long trunk piston, the front end of which serves as a guide and does not enter the cylinder proper ; the connecting rod communicates between the guide and the crank shaft, the side thrust is thus kept off the piston and cylinder proper, which become hot. The crank shaft is heavy and the fly-wheel a large one ; considerable energy being required to take the piston through the negative part of the cycle. The cylinder is considerably longer than the piston stroke, so that the piston when full in leaves a considerable space into which it does not enter.

Outside the cylinder, running across it at the end of the space, works a large slide valve ; it is held against the cylinder face by a cover plate and strong spiral springs ; it is driven to and fro by a small crank, on the end of a shaft parallel to the cylinder axis,

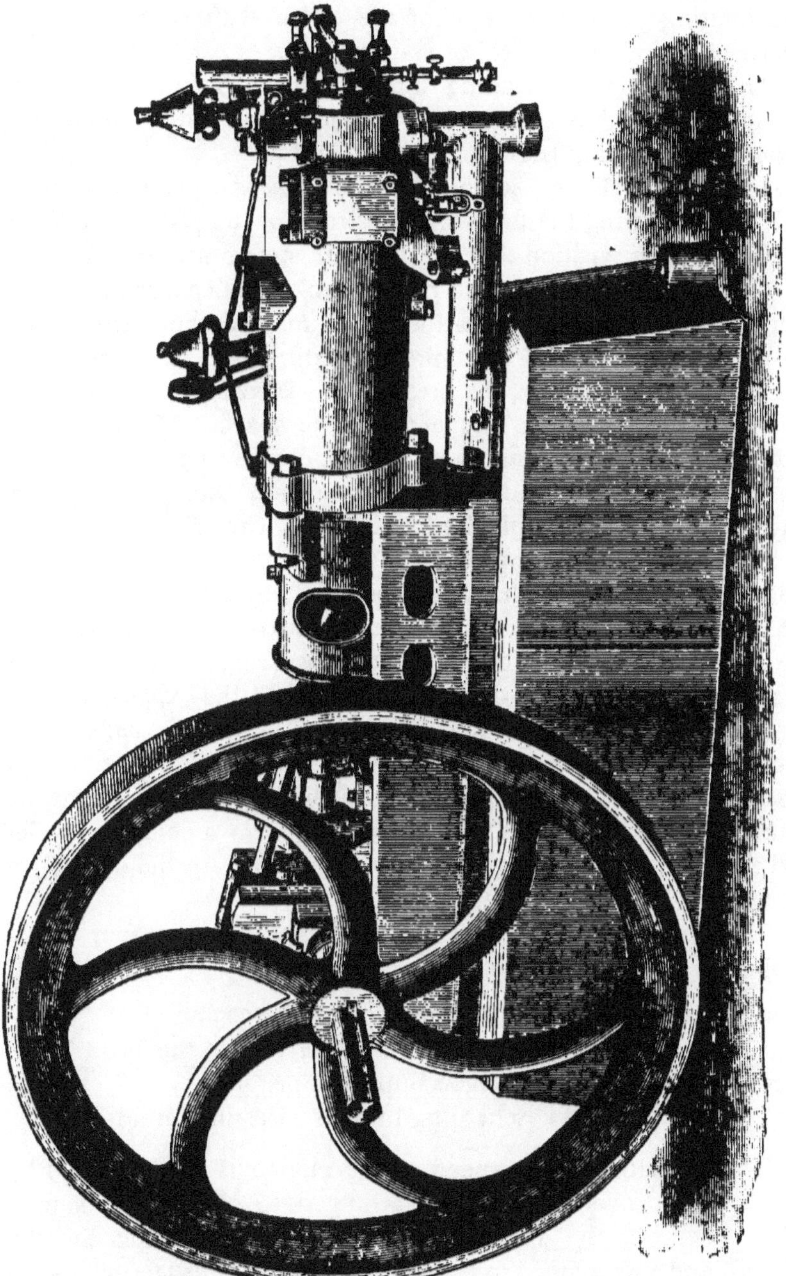

Fig. 50.—Otto Engine.

and rotating at half the rate of the crank shaft, from which it receives its motion by bevel or skew gearing.

An exhaust valve, leading into the space by a port, is also actuated at suitable times from the secondary shaft; so are the governing and oiling gear.

The single cylinder serves alternately the purposes of motor and pump; during the first forward stroke of the piston, the slide valve is in such position that gas and air stream into the cylinder from the beginning to the end of the stroke, the charge mixing as it enters with whatever gases the space may contain; the return stroke then compresses the uniform mixture into the space, and when the piston is full in, the pressure has increased to an amount determined by the relative capacity of the space. Meantime the slide valve has moved to another position, first closing the admission gas and air ports, to permit of the compression, then bringing on a cavity in the valve which is filled with flame, when the compression is completed. The compressed charge therefore ignites and the pressure rises so rapidly that maximum is attained before the piston has moved appreciably on its forward stroke (second stroke); the piston is thus under the highest pressure at the beginning of its stroke and the whole stroke is available for the expansion.

This is the motive stroke. At the end of it, the exhaust valve opens and the return stroke is occupied in driving out the burned gases, except that portion remaining in the space which cannot be entered by the piston. These operations form a complete cycle, and the piston is again in the position to take in the charge required for the next impulse.

The cycle requires two complete revolutions, or four single strokes.

First out stroke.   Charging cylinder with gas and air.
,, in ,,   Compressing the charge into the space.
Second out stroke.   Explosion impelling piston.
,, in ,,   Discharging burned gases into atmosphere.

The regulation of the speed of the engine is accomplished by a centrifugal governor, which is arranged to close a gas supply valve whenever the speed increases. An explosion is thereby missed, and the engine goes through its cycle as usual, but as no

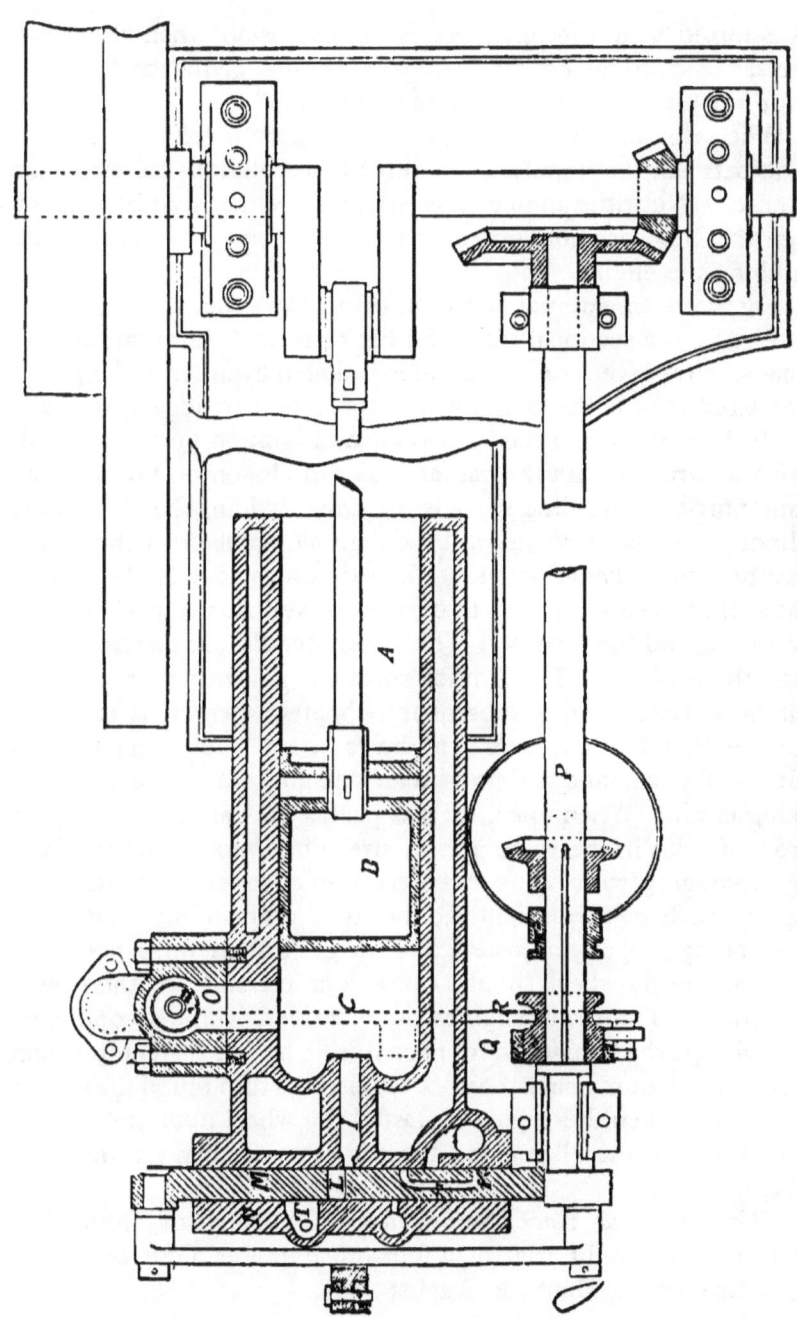

FIG. 51.—Otto Engine (Sectional plan).

gas is mixed with the air, there is no explosion when the flame enters, the compressed air merely expanding, giving back to the piston the energy taken during compression.

When running without load, 8 or even more revolutions may be made between the impulses, at full load 2 revolutions are made per impulse. Notwithstanding this irregularity the fly-wheel is so large that no variation observab'e by the eye can be seen while watching the engine.

Fig. 50 is an external elevation of an Otto engine.

Fig. 51 is a sectional plan, and fig. 52 an end elevation showing exhaust valve lever. A is the water-jacketed cylinder, B the piston shown full in, C is the compression space or cartridge space as it is called by Million; I the admission and ignition port, communicating alternately with the gas and air admission port K, and the flame port L in the slide M; N is the cover holding the slide to the cylinder face and carrying in it the external flame for lighting the movable one in flame port L. The exhaust valve is of the conical seated lift type and is seen at O; it is driven from the shaft P by the cam Q and the lever R. The other details are clearly shown upon the drawing. The ignition valve and governing arrangement will be described in a subsequent chapter; here it is sufficient to state that the governor withdraws a cam actuating the gas valve S, fig. 52, and so prevents it opening when the piston is taking in air. When open, the gas passes the valve, then through a row of holes in the valve port K, streaming into the air and mixing thoroughly with it as it enters the cylinder. To start the engine, the flame at T is lighted; the cock commanding the internal flame being properly adjusted, and the gas turned on, a couple of turns at the fly-wheel should cause ignition and set the engine in motion. The larger engines are provided with a second cam, which keeps the exhaust valve open during half of the compression stroke and so diminishes the work required to turn round the engine by hand. When the engine is started the wheel upon the lever is shifted to the normal cam and the compression then returns to its usual intensity.

Diagrams and Gas Consumption.—Dr. Slaby, of Berlin, has made a very careful trial of a four-horse power Otto engine at Mr. Otto's works, Deutz, in August 1881.

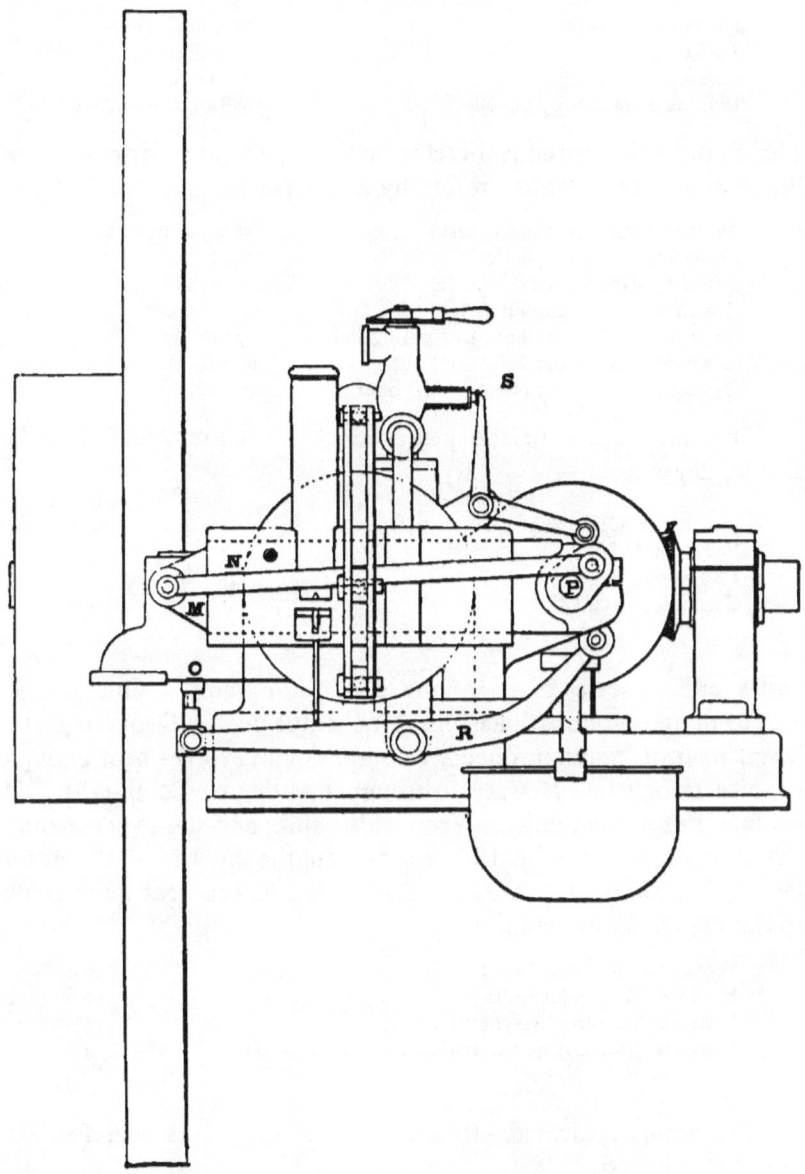

FIG. 52.—Otto Engine (End elevation).

The dimensions of the engine are:

Diameter of cylinder	171·9 millimetres.
Stroke	340 millimetres.
Compression space	4770 cb. centimetres.
Volume displaced by piston	7888 cb. centimetres.

The compression space is therefore 0·6 of the volume displaced by the piston. The results are briefly as follows:

Average revolutions during test	156·7 per minute.
Power indicated in cylinder	5·04 horse.
Power by dynamometer	4·4 horse.
Gas consumed in one hour	142·67 cb. ft.
Gas consumed in one hour by igniting flames	2·73 cb. ft.
Gas consumption per IHP per hour	28·3 cb. ft.
Gas consumption per effective HP hour	32·4 cb. ft.

The composition of the gas used at the Gasmotoren-Fabrik, Deutz, is given as—

	Volumes.
Marsh gas, CH_4	34·4
Ethylene, C_2H_4	3·5
Hydrogen, H	56·9
Carbonic oxide, CO	5·2
	100·0

and 1 cubic metre of it weighs 0·404 kilograms. One pound weight of it therefore measures 39·6 cubic feet. Deducting the latent heat of steam produced, 1 pound weight evolves heat enough to raise 12,094 lbs. of water, through one degree Centigrade. It evolves 12,094 heat units. From this value, and the experimental determination of the heat leaving the engine by way of the water jacket, Dr. Slaby calculates the disposition of 100 heat units given to the engine as follows:

Work indicated in cylinder	16·0
Heat lost to cylinder walls	51·0
Heat carried away by exhaust	31·0
Heat lost from engine by conduction and radiation	2·0
	100·0

The actual indicated efficiency of the engine is therefore 16 per cent. or 0·16.

The temperature of the gases expelled during the exhaust stroke was determined by carefully protecting the exhaust pipe from loss of heat by non-conducting material, and then seeing whether zinc or antimony would melt in it. Zinc melted but antimony did not; as the melting point of zinc is 423° C., and the antimony melting point is 432° C., the temperature of the exhaust gases is given with great accuracy as between these two temperatures. The average composition of the mixture is given as 1 vol. coal gas to 13·73 vols. of air and other gases. Here Dr. Slaby is plainly in error, as his own figures conclusively show. The volume of coal gas taken into the engine at each stroke as measured by the gas meter is given as 859 cubic centimetres, the total volume swept by the piston of the engine per stroke is 7888 cubic centimetres, the volume of the compression space 4770 cubic centimetres. Now if the gas be introduced into the cylinder while it is filled completely, space included, with cold gases, at the same temperature as the gases when measured by the meter, this figure is correct enough. But the gases are not so introduced, the space is already filled with exhaust gases at a temperature of about 400° C. by Dr. Slaby's own determination; this volume must therefore be calculated to atmospheric temperature before an approach to the true ratio can be obtained. Taking atmospheric temperature at 17° C., then 4770 cubic centimetres of burned gases at 400° C. becomes reduced to 2055 cubic centimetres at 17° C.; that is, the total charge will consist of 859 cubic centimetres of coal gas, 7029 cubic centimetres of air, and 2055 cubic centimetres of burned gases from the previous explosion.

The ratio is

$$\frac{\text{coal gas}}{\text{air and burned gases}} \quad \frac{859}{7029 + 2055} = \frac{1}{10\cdot 5}$$

The composition of the charge is more correctly represented as 1 vol. of gas to 10·5 vols. of air and other gases. Even here, however, the dilution is overstated, as it is assumed that the piston has taken in the charge at full atmospheric temperature and pressure. But there is some throttling in passing through the admission valve and port, and also some heating of the air by striking the piston and cylinder walls. Professor Thurston, in experiments to be described later on, proves this to be the case, and shows that the charge is even stronger than has been calculated.

It has been already proved that in this type of engine, expanding after compression and explosion to the same volume as existed before compression, the theoretic efficiency is independent of the temperature of the explosion or the temperature existing before compression, and depends only upon the volume before and after complete compression. As the ratio of compression space to volume swept by the piston is 0·6 to 1, the volume before compression is 1·6, volume after compression 0·6.

The theoretic efficiency is (p. 53) $E = 1 - \left(\dfrac{v'_c}{v'_o}\right)^{\gamma-1}$, and v'_c is the compression volume, and v'_o the volume before compression; in this case $E = 1 - \left(\dfrac{0\cdot6}{1\cdot6}\right)^{\cdot408}$ or $1 - \left(\dfrac{1}{2\cdot66}\right)^{\cdot408}$; here $E = 0\cdot33$.

That is, if all the heat were given to the engine at the moment of complete explosion at the beginning of the stroke, and no heat were lost to the cylinder during the expansion to the original volume, then 33 per cent. of that heat would be converted into indicated work. But the author's explosion experiments give the factor necessary for correcting this theoretical number (p. 113). Taking the mixture of 1 gas to 10 vols. air as nearest, the efficiency of the gas in it is 0·46; that is, during the time of the forward stroke, taken as 0·2 sec., 1 vol. of gas is required to produce and keep up a pressure which 0·46 vol. would suffice for if it was all applied to heating and no loss by cooling.

The actual indicated efficiency of the engine using this mixture and this expansion and compression should be 0·33 × 0·46 = 0·152 nearly. That is, the engine should convert 15·2 per cent. of the heat given to it into work. Dr. Slaby's number, found by experiment, is 16 per cent. The numbers are exceedingly close.

The mechanical efficiency of the engine is high, the ratio of dynamometric to indicated power being 87 to 100, and the friction of the engine only 0·64 horse

Professor Thurston's Experiments on a 6 HP Otto Engine.

Dr. Slaby's experiments are exceedingly complete, but Professor Thurston in America has made even more extended measurements.

Messrs. Brooks and Steward made the trials under the direction of Professor Thurston, at the Stevens Institute of Technology, Hoboken. The dimensions of the engine are as follows:

Diameter of cylinder	8·5 ins.
Stroke	14 ins.

Capacity of compression space 38 per cent. of total cylinder volume.

Not only was the gas entering the engine measured, but at the same time the air required was measured through a 300 light meter. So far as the author is aware, this is the only set of experiments in which this was done; it is by far the most accurate way of getting the true proportions of the explosive mixture.

The temperature of the exhaust was measured by a pyrometer, and the power determined, both by indicator and dynamometer; at the same time the heat passing into the walls of the cylinder was determined by measuring the water heated and estimating the loss by radiation and conduction.

The total number of revolutions during the various tests were taken by a counter. Many trials were made under varying conditions of load and mixture used. The following is the best full-power trial, giving the most economical results:

Average revolutions during test	158 per minute.
Power indicated in cylinder	9·6 horse.
Power by dynamometer	8·1 horse.
Gas consumed in one hour	235 cb. ft.
Gas consumption per IHP per hour	24·5 cb. ft.
Gas consumption per effective HP per hour	29·1 cb. ft.

An analysis of the gas used during the trials made by Thomas B. Stillman, Ph.D., is as follows:

Hydrogen, H	39·5
Marsh gas, CH_4	37·3
Nitrogen, N	8·2
Heavy hydrocarbons, C_2H_6, &c.	6·6
Carbonic oxide, CO	4·3
Oxygen, O	1·4
Water vapour and impurities (H_2O, CO_2, H_2S)	2·7
	100·0

One cubic metre of this gas weighs 0·606 kilograms. One pound weight of it therefore measures 26·43 cubic feet. One pound when completely burned evolves heat enough to raise 9070 lbs. water through 1° C.

The air necessary to supply just enough oxygen for the complete combustion of 1 vol. of this gas is 5·94 vols.

From these values and experiments upon temperature of the exhaust gases, Professor Thurston estimates the disposition of 100 heat units by the engine as follows:

Work indicated in cylinder	17·0
Heat lost to cylinder walls	52·0
Heat carried away by exhaust gases	15·5
Heat lost from engine by conduction and radiation	15·5
	100·0

The actual indicated efficiency is therefore 17 per cent.

The number showing the proportion of heat passing into the water jacket is also very nearly Slaby's, but the amount expelled with the exhaust is much understated. The amount lost by radiation is overstated.

The temperature of the exhaust gases, as determined by a pyrometer placed in the exhaust pipe, varies in the experiments at full load from 399° C. to 432° C., thus practically coinciding with Slaby. The ratio of air to gas was found, by actual measurement of both, to be about 7 to 1 when the engine was working most economically. Although with better gas the ratio would be slightly increased, yet it could not equal that usually given for the Otto engine, 10 to 1 or thereabouts.

The ratio is commonly obtained from a measurement of the gas consumption alone, the air being reckoned as the volume of the piston displacement, less the measured amount of gas. This is not an accurate method, for the reason already stated.

Gas Engines of Different Types in Practice. 177

If the mixture filling the cylinder mingles with the burned gases filling the compression space, then the average composition of the charge is 1 vol. coal gas to 9·1 vols. of other gases.

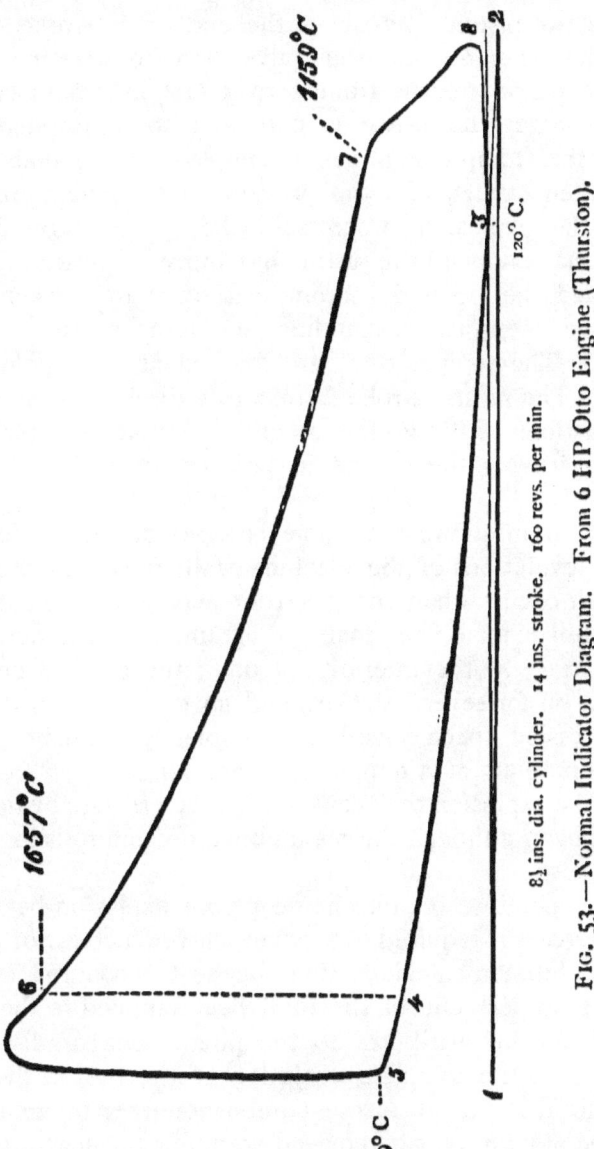

Fig. 53.—Normal Indicator Diagram. From 6 HP Otto Engine (Thurston). 8½ ins. dia. cylinder. 14 ins. stroke. 160 revs. per min.

Fig. 53 is a fair sample of the diagrams obtained during Professor Thurston's tests while the engine was giving full power. The piston while moving from the point 1 to the point 2 takes in the charge; the pressure in the cylinder falls below atmosphere as the piston approaches the end of its stroke. This is due to the resistance of the valve port to entering air and gas. The piston returns from 2 to 5 (1st in-stroke) compressing the charge, the pressure increasing to atmosphere at the point 3, the compression being complete at the point 5; the ignition then occurs, and the pressure and temperature rapidly rises as the explosion progresses; the temperature does not attain its maximum till the piston has moved forward a little and has reached the point 6. From that point to 7, when the exhaust valve opens, the expanding line is as nearly as possible adiabatic. The temperatures are marked at each point of the diagram. The return stroke from 2 to 1 discharges the products of combustion. This is the second in-stroke, completing the cycle and leaving the engine in position to again take in the charge.

The diagram shows the whole changes occurring during two complete revolutions of the machine while fully loaded. Fig. 54 shows what occurs when the governor acts, when the engine is at less than full load. The smaller diagram, B, is the normal one, and the larger, A, the intermittent one; the gas has been completely cut off for several strokes, and so the hot burned gases in the compression space have been completely discharged and replaced by pure air at a temperature not far removed from atmospheric; the explosion then causes a higher pressure by nearly half an atmosphere, although the maximum temperature is less than in the usual case.

The temperature of the charge before explosion being less, a smaller increase is required to produce a given increase of pressure. Professor Thurston calculates that the heat accounted for by the diagram is 60 per cent. of the total heat supplied to the engine; the deficiency he attributes to the phenomena of dissociation, which prevents the complete evolution of the heat at the highest temperature, but permits further combustion when the temperature falls. The amount of gas required to run at full speed, 166 revolutions per minute without any load, was found to be from 50 to 70 cubic feet per hour.

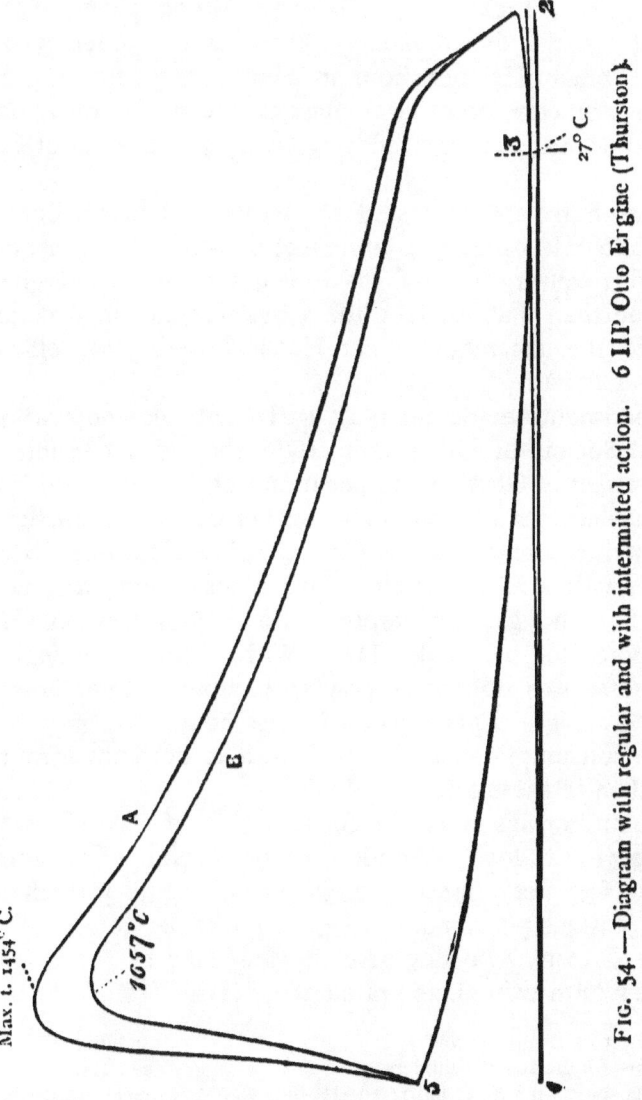

Fig. 54.—Diagram with regular and with intermitted action. 6 HP Otto Engine (Thurston).

Other tests of Otto Engine.—The experiments of Dr. Slaby and Professor Thurston upon the Otto engine are by far the most complete which have yet been made to the author's knowledge.

Some tests given in Schöttler, however, will be quoted. A four HP engine was found to consume as a best result 32·4 cubic feet of gas per brake HP per hour in Altona, giving at the time 3·96 HP on the dynamometer. Another consumed 33·7 cubic feet per brake HP per hour in Hanover, giving 4·95 HP on the dynamometer; to drive the last engine at 160 revolutions per minute without load required 41·3 to 43·4 cubic feet of Hanover gas.

A two-horse engine, tested by Brauer and Slaby, Berlin, gave 2·28 brake HP, using 35·3 cubic feet per brake HP per hour.

In this country the coal gas in common use is of higher heating value than that used on the Continent and in America; accordingly the gas required per HP is less, but the efficiency is almost identical.

Experiments made upon an 8 HP Otto engine by the Philosophical Society of Glasgow in 1880, showed a consumption of 22 cubic feet of Glasgow gas per indicated HP, giving 9 HP upon the dynamometer, and 28 cubic feet per dynomemetric horse.

Experiments made at the Crystal Palace Electrical Exhibition, in 1881, with a 12 HP engine gave a maximum brake power of 18·3 HP, with a gas consumption of 23·7 cubic feet per IHP, and 29·1 cubic feet per brake HP. With a two-horse engine, 2·87 brake horse was obtained upon 33·4 cubic feet per horse hour, and 27·9 cubic feet per indicated horse hour.

The consumption running without load does not seem to have been taken in these tests.

The author has taken the consumption of a two-horse engine running without load in London, at 160 revolutions per minute, as 32 cubic feet per hour, and a 3·5 horse engine without load at 166 revolutions per minute as 43 cubic feet per hour.

The Messrs. Crossley give the following as the results with their new Otto twin engine rated at 12 HP :

Power by dynamometer	23 horse.
Power indicated in cylinders	28 horse.
Gas consumption per indicated HP	20 cb. ft. per hour.
Gas consumption per effective HP	24·3 cb. ft per hour.
Total consumption at full power	560 cb. ft. per hour.
Total consumption when running without load at 160 revs. per minute	100 cb. ft. per hour.

Gas Engines of Different Types in Practice 181

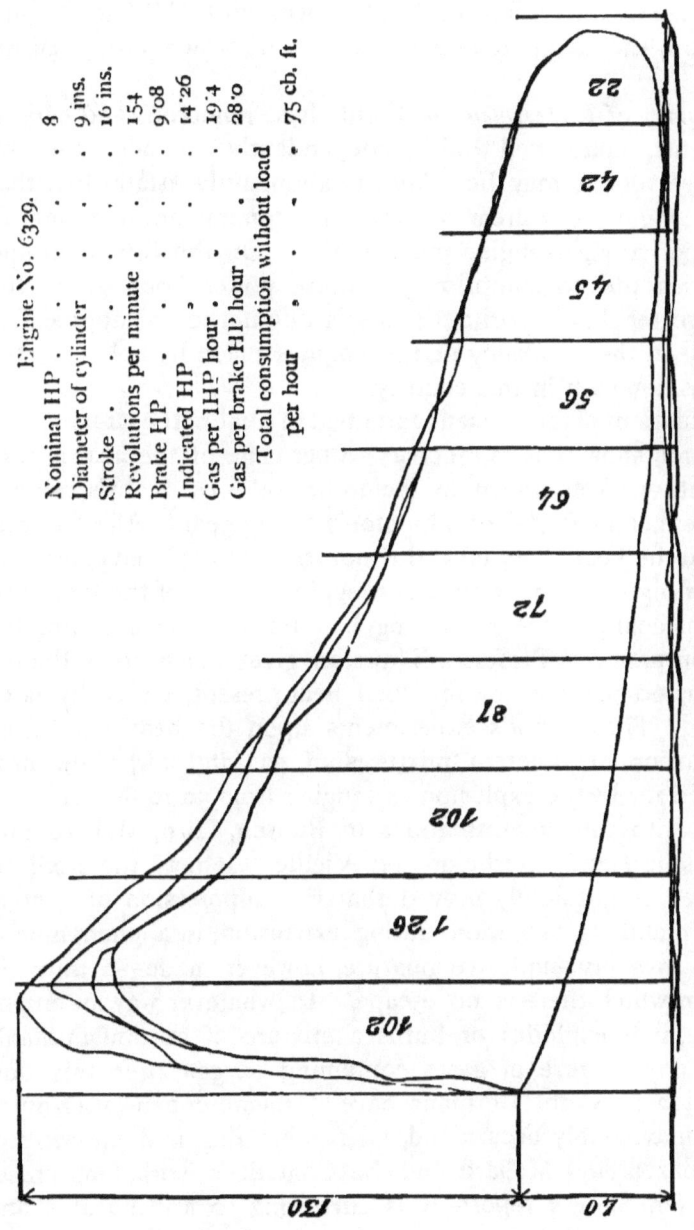

Fig. 55.—Diagram from 8 IIP Otto Engine (Garrett).

Engine No. 6329.
Nominal HP 8
Diameter of cylinder . . . . 9 ins.
Stroke . . . . 16 ins.
Revolutions per minute . . . 154
Brake HP . . . . . . 9·08
Indicated HP . . . . . 14·26
Gas per 1HP hour . . . . 19·4
Gas per brake HP hour . . . 28·0
Total consumption without load per hour . . . . . 75 cb. ft.

These results are obtained using Manchester gas.

Mr. G. H. Garrett has made a trial with an 8 HP Otto engine in Glasgow, the diagram and particulars of which are given on fig. 55.

Summary of Experiments.—From these numerous and careful experiments, conducted quite independently of each other by many observers, it may be taken as abundantly established that the Otto engine is a great advance in economy and certainty of action upon any gas engine preceding it. On the Continent and in America the consumption per horse power hour is, on the whole, greater than in Britain; this is due not to any appreciable difference in the efficiency of the engines made here, but to the better gas common in this country.

Calculations of the efficiency attained in some of the later engines in England, show that as much as 18 per cent. of the heat is converted into work as shown by the indicator. Dr. Slaby's value is 16 per cent., and Professor Thurston's 17 per cent. All observers agree that the heat liberated at the moment of completed explosion, that is, of highest temperature, is roughly one-half of the total heat present as coal gas, the remaining half being evolved during the expansion period. Professor Thurston gives the heat of the explosion as 60 per cent. of the total heat present, Dr. Slaby as 55 per cent. The author's experiments upon the heat evolved by the explosion of different mixtures of gas and air, show heat accounted for by the explosion as ranging from 50 to 60 per cent., agreeing with the determinations of Bunsen, Hirn, Mallard and Le Chatelier, and Berthelot and Vieille. It may therefore be considered as absolutely proved that this suppression of heat at explosion, and its evolution during expansion, is a phenomenon inherent in every explosive mixture, however made—a thing, in fact, from which there is no escape. In whatever way an engine be made, if it explodes or burns a mixture of any inflammable gas with any mixture of gases containing oxygen, then this slow combustion or, as the Germans have it, nachbrennen (after-burning) is unavoidably occasioned. Knowing this, and knowing of Hirn, Bunsen, and Mallard and Le Chatelier's work long precedent to Dr. Slaby's report, it is surprising to find so able and learned a scientist quoted as stating that in the Lenoir engine the whole heat was evolved at the moment of complete explosion. In the Lenoir, as in every other gas engine which has ever been constructed, not more than one-half of the whole heat of the gas present is then evolved, the remaining heat being evolved on the expanding stroke.

Schöttler falls into the same error, and, although mentioning Wedding's statement of Bunsen's law of dissociation, shows that he rejects it when he assumes that the whole heat is evolved. A very cursory examination of the Lenoir diagram would at once prove to Prof. Schöttler that Lenoir did not succeed in so escaping the laws of nature; had he done so, there would have been no necessity for our modern improvements.

The consumption of continental gas may be taken as varying between 32 cb. ft. and 35 cb. ft. per effective HP per hour, and about 28 cb. ft. per IHP per hour.

In Britain it may be taken as ranging from 24 cb. ft. per effective HP to 33 cb. ft., and 20 to 24 cb. ft. per IHP per hour, depending upon the quality of gas used and on the dimensions of the engine tested. Other things being equal, better results are obtained with large engines. The theoretic efficiency is constant for both large and small engines where the same compression is in use, but the loss of heat from the explosion to the sides of the cylinder is less in the large engines, due to the diminished surface exposed in proportion to the volume used. The effect is to increase the efficiency of the gas in the mixture used, a smaller quantity being necessary to make up for the loss of heat.

The indicator diagrams prove the very efficient nature of the Otto cycle. The great simplicity attained by the alternate use of the cylinder as pump and motor diminishes the number of valves necessary, and secures the minimum resistance to the entering gases, while entirely preventing any loss due to ports, in transferring the gases from one cylinder to another. The carrying out of the cycle is mechanically almost perfect, no work being spent which is not included in the theory. Again, the piston is full in at the moment of ignition and is almost at rest; the heat, producing maximum temperature, is therefore added at nearly constant volume. The highest pressure which the gas present is capable of producing is therefore attained at the beginning of the stroke simultaneously with the highest temperature; the succeeding expansion is then very rapid, and so no unnecessary waste of heat occurs, the temperature being rapidly depressed by work being done. The united effect of all the arrangements is seen in a diagram which is almost theoretically perfect; the only deduction from theory is due to the unfortunate property of explosive mixtures of continued combustion after explosion. And this reduces the theoretic efficiency to one-half in practice. The theoretic efficiency of all Otto engines, of whatever dimension, is 0·33, as the compression space in all cases bears nearly the ratio of 0·6 to 1·0 when compared with the cylinder volume which is swept by the piston. The actual indicated efficiency is very nearly one-half of that number.

If combustion by any means could be made complete at the highest temperature and pressure at the beginning of the stroke, instead of continuing as it does well into the expansion stroke, then greatly increased economy would result, and in large engines theory might be very nearly approached.

This point will receive further discussion later on.

Clerk Engine.—Otto's method is probably the readiest and easiest solution of the problem of attaining in a practicable manner the advantages of compression; in some points, however, the advantages are accompanied with compensating disadvantages.

Only one impulse for every two revolutions is obtained; the engine is therefore stronger and heavier than need be if impulse every revolution were possible. It is also more irregular in its action than more frequent impulses would give.

The Clerk engine was invented by the author with the view of obtaining impulse at every revolution, while getting at the same time the economy due to compression.

At first blush it seems a very simple matter to make a compression gas engine to give an impulse for every revolution; this was the author's opinion when he commenced work for the first time upon gas engines using compression in October 1876. Since then he has had occasion to modify the opinion: the difficulties are very great; any engineer who doubts this will speedily be convinced upon making the attempt.

Gas Engines of Different Types in Practice 185

It was not till the end of 1880 that the author succeeded in producing the present Clerk engine; before that time he had several experimental engines under trial, one of which was exhibited at the Royal Agricultural Society's show at Kilburn in July 1879. This engine was identical with the Lenoir in idea, but with separate compression and a novel system of ignition.

The Clerk engine at present in the market was the first to succeed in introducing compression of this type, combined with

FIG. 56.—The Clerk Gas Engine.

ignition at every revolution; many attempts had previously been made by other inventors, including Mr. Otto and the Messrs. Crossley, but all had failed in producing a marketable engine. It is only recently that the Messrs. Crossley have made the Otto engine in its twin form and so succeeded in getting impulse at every turn.

In the Clerk engine the whole cycle is completed in one revolution, and an impulse given to the crank on every forward stroke of the piston, when working at full power.

The engine contains two cylinders, one for producing power,

the other for taking in the combustible charge and transferring it to the power cylinder. At the end of the motor cylinder is left a compression space of a conical shape, and communicating with the charging or displacing cylinder by a large automatic lift-valve opening into the space; at the other end of the cylinder are placed V-shaped ports opening to the atmosphere by the exhaust pipe; the motor piston, when near its outer limit, overruns these ports and allows the cylinder to discharge. The pistons are connected in the usual manner by connecting rods, the motor to the main crank of the engine, the displacer to a crank pin in one of the arms of the fly-wheel; the displacer crank is in advance of the motor crank, in the direction of motion of the engine, by a right angle. The displacer piston on its forward movement takes in its charge of gas and air, and has returned a fraction of its stroke when the motor piston uncovers the exhaust ports. While crossing the centre, opening and shutting these ports the displacer piston has moved in almost to the end of its cylinder, discharging its contents into the space and forcing out at the exhaust ports the products of the previous ignition. The proportions of the two cylinders are so arranged that the exhaust is as completely as possible expelled, and replaced by cool explosive mixture, which thoroughly mixes with any exhaust remaining, cooling it also to a considerable extent. Care must be taken in the arrangement of the parts that an excessive volume is not sent from the displacer, otherwise it may reach the exhaust ports and gas discharge unburned.

The return stroke of the motor piston now compresses the mixed gases, and when at the extreme end, the igniting valve fires the mixture, the piston moves forward under the pressure thereby produced, till the opening of the exhaust ports causes discharge and replacement as before. In this way an impulse is given at every revolution, and the motive power applied to greater advantage. The motor cylinder is surrounded by water for cooling, but this is unnecessary with the displacer, as it uses only cool gases. The pressures used are high, so that both motor piston and its connections are made very strong; the pressure on the displacer piston is very little, so the connections are light. It is not a compressing pump, and is not intended to compress before introduction into the motor, but merely to exercise force enough to pass the gases through the lift valve into the motor cylinder, and there displace the burnt gases, discharging them into the exhaust pipe. The pressure to be overcome is only that due to resistance in the exhaust pipes and the lift valve.

The inlet valve for gas and air is also automatic; its seat is of the usual conical kind but somewhat broad. A gutter runs round the centre, having small holes bored through to a recess behind,

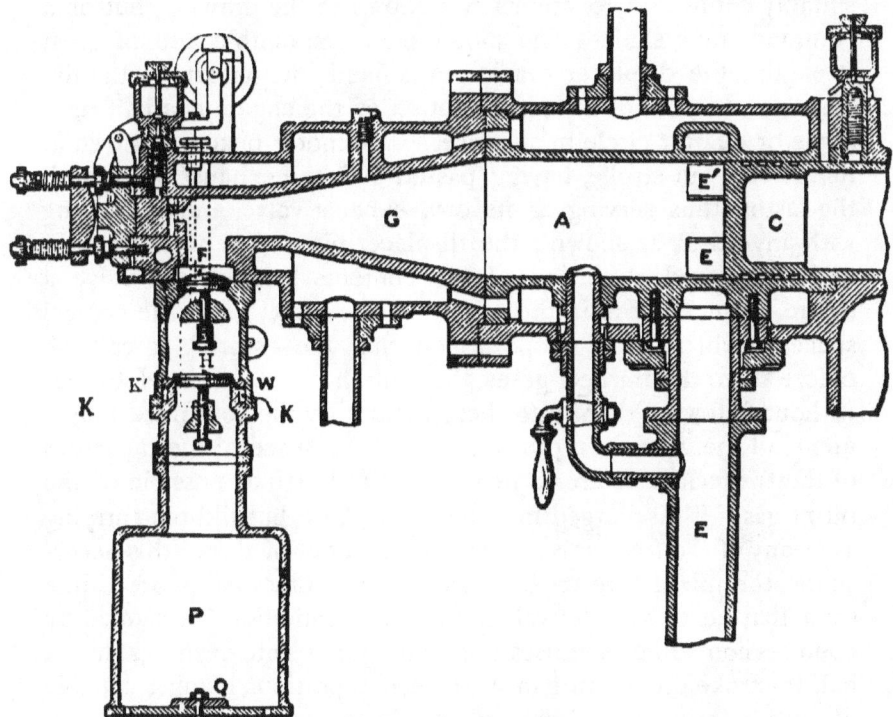

FIG. 57.—Longitudinal Section of Clerk Gas Engine.

which communicates with the gas supply pipe. The suction lifts the valve to a certain height, and, as the gases enter, the air flows past the holes and becomes thoroughly impregnated with gas, the extent being determined by the number of holes and the proportion of their area to the total area of the valve opening. The upper valve is made heavy to withstand the maximum pressure of the explosion; both valves are arranged so that the guide forms a piston working in an air cylinder, so arranged as to check the fall of the valve before touching the seat, and so prevent any disagreeable rattle.

Description of the Drawings.—Fig. 56 is a general view of the engine. Fig. 57 a longitudinal section. Fig. 58 a sectional plan. In these drawings all the essential parts of the engine are represented; the sectional plan (fig. 58) shows the two cylinders, motor A and displacer B, in which work the pistons C and D suitably connected to cranks not shown in the drawing, but on a common crank shaft. The motor crank is double and of great strength; the displacer crank pin is fixed into an arm in the flywheel, and in the direction of motion of the engine is a half right angle or quarter circle in advance. The motor piston is shown at its extreme out-stroke, having passed over the exhaust ports $E\ E^1$, the piston thus serving as its own exhaust valve, and dispensing with any other, as shown; the displacer piston has moved half in and discharged a portion of the contents through the valve F (more distinctly seen in the other section, fig. 57) into the conical space G, which is so proportioned that the entering gases push before them the burned gases through the ports referred to, but without following them into these ports. By the continued movement, all the gases in B pass into A and the space G; the capacities of the two cylinders are so related that as much as possible of the burnt gases is discharged into the atmosphere, but without carrying away any of the fresh mixture containing unburned gas; this necessitates the mixture next the piston being somewhat more dilute than that next the inlet valve, but the commotion occasioned by compression so far equalises this undesirable state of things that at half in-stroke the mixture in its weakened portions is quite capable of inflammation by a light or the electric spark. The piston D having completed its in-stroke, C has passed over the discharge ports and compresses the contents of the cylinder into the space G; when full in and therefore completely compressed, the slide valve M has moved into such a position as to ignite the mixture; the maximum pressure being attained very rapidly and before the piston can move appreciably on its out-stroke, the piston is impelled forward under the pressure produced until it reaches the ports $E\ E^1$, when the contents are rapidly discharged, and the interior and

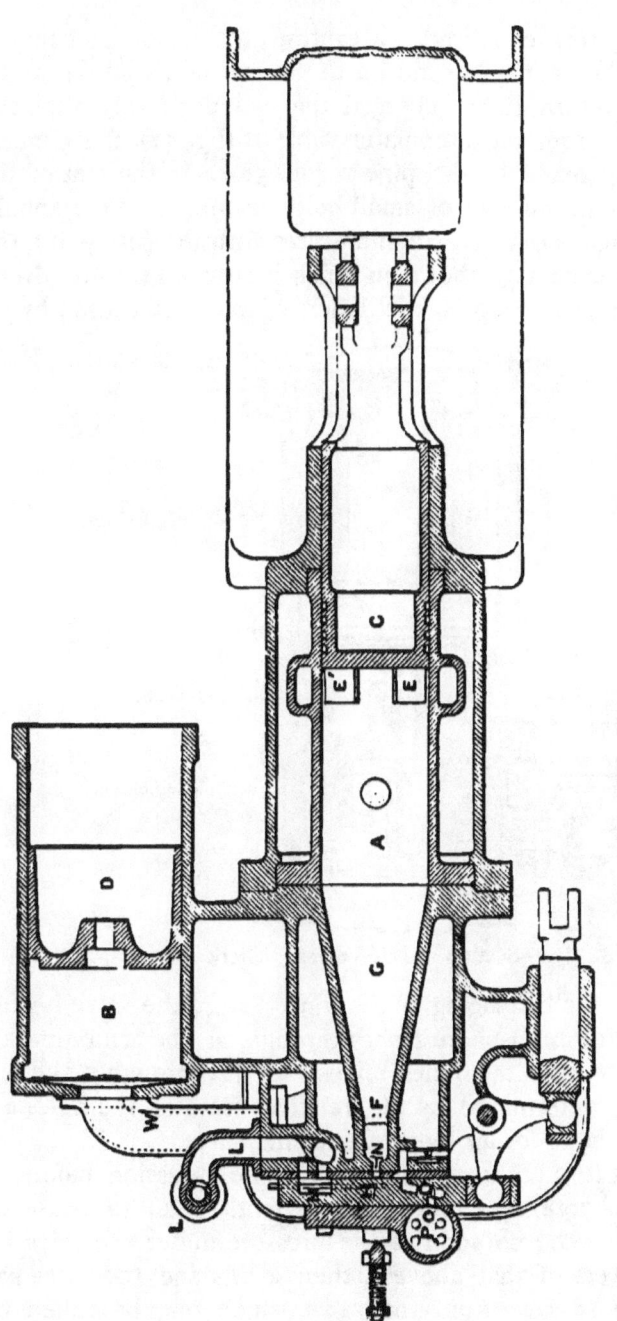

Fig. 58 Sectional Plan of Clerk Gas Engine.

exterior pressures equalised. Meantime the piston D being in advance of the motor has moved to the end of its stroke and is beginning to return, it has charged the cylinder B with a mixture of gas and air from the automatic valve H (fig. 57), the communication being made by the pipe W (fig. 58). In the seat of this valve are bored a number of small holes passing into the annular space K K¹ (fig. 57), which communicates with the gas cock L (fig. 58) through the passage shown, in which is situated the lift-governing valve, not seen. Under the deficit of pressure caused by the

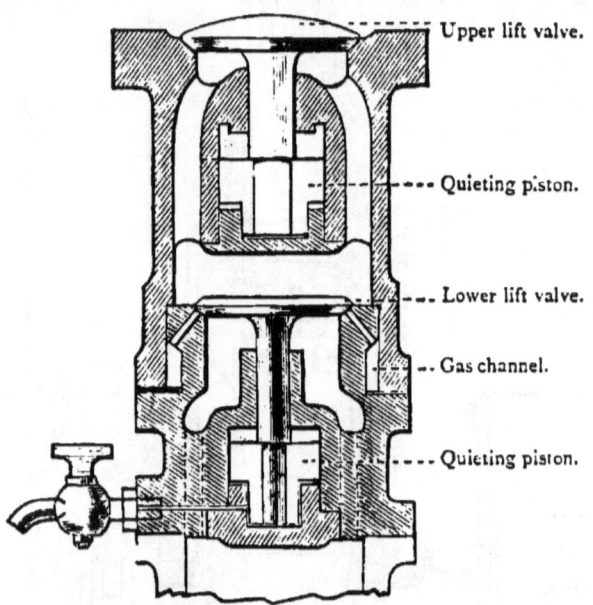

Fig. 59.—Section of Lift Valves. Clerk Engine.

movement of the displacing or charging piston, the valve is lifted and the exterior atmosphere rushes through, at the same time the gas passing through the holes mixes with it thoroughly, the proportion being determined by the relative areas of the holes and the space available for air by the lift of the valve.

The gases in B are under some slight compression before the complete discharge of A, but not sufficiently great to cause any material resistance; so soon as the pressure under the valve F is slightly in excess of that above it, then it lifts and the gases pass into G. The passage from the valve, which may be called the upper lift valve, is more clearly seen in fig. 57 : the igniting hole is shown at N, and communicates at the proper time with flame in the cavity O, which has been ignited at the exterior flame P, from a Bunsen burner (fig. 58).

The two automatic valves charging the displacer cylinder and discharging into the motor cylinder are provided with quieting pistons, cushioning the blow on the valve seat and preventing rattle; they are similar to the dash pot contrivances used on Corliss' steam engines to check the snap of the steam valves, but, unlike them, are attached directly to the valve, instead of to the valve spindle and guide. The arrangement is very clearly seen at fig. 59: the lower valve has no spring, it returns to its seat by its own weight; but the upper valve requires to act more quickly and is pulled down by a spring.

The piston attached compresses the air before it, and the valve strikes its seat rapidly but without jar or recoil.

The igniting slide, M, is driven from an eccentric on the crank shaft through a bell crank and guide.

Diagrams and Gas Consumption.—The following tests give the latest results from the Clerk engine; they are the usual trials

TESTS OF THE CLERK ENGINES OF VARIOUS POWERS.

	2 HP	4 HP	6 HP	8 HP	12 HP
Diameter of motor cylinder	5 ins.	6 ins.	7 ins.	8 ins.	9 ins.
Stroke	8 ins.	10 ins.	12 ins.	16 ins.	20 ins.
Diameter of displacer cylinder	6 ins.	7 ins.	7½ ins.	10 ins.	10 ins.
Stroke	9 ins.	11 ins.	12 ins.	13 ins.	20 ins.
Average revs. per min. during test	212	190	146	142	132
Average pressure (available) in motor cylinder in lbs. per sq. in.	43·2	63·9	53·2	60·3	64·8
Power indicated in motor cylinder	3·62	8·68	9·05	17·38	27·46
Power by dynamometer	2·70	5·63	7·23	13·69	23·21
Gas consumption in cb. ft. per IHP per hour	29·8	24·19	24·3	20·94	20·39
Gas consumption per brake HP hour	40·0	37·3	30·42	26·58	24·12
Max. pressure of explosion in lbs. per sq. in. above atmos.	155 lbs.	235	195	195	238
Pressure of compression in lbs. per sq. in. above atmos.	33 lbs.	55	48	49	57
Displacer resistance	0·40	0·80	0·86	1·50	2·00
Gas consumed per hour by each engine at speed without load	40 cb. ft.	58 cb. ft.	57 cb. ft.	70 cb. ft.	90 cb. ft.

made by Messrs. L. Sterne and Co. on all engines before leaving the works, and therefore represent fairly the economy to be expected from these engines in ordinary work. They are from 2, 4, 6, 8, and 12 HP engines (nominal). The trials were made during 1885 at the Crown Iron Works, Glasgow, under the direction of Mr. G. H. Garrett.

Figs. 60, 61, 62 are fair samples of the diagrams taken during the tests. Figs. 63 and 64 are diagrams from the displacers showing the displacer resistance.

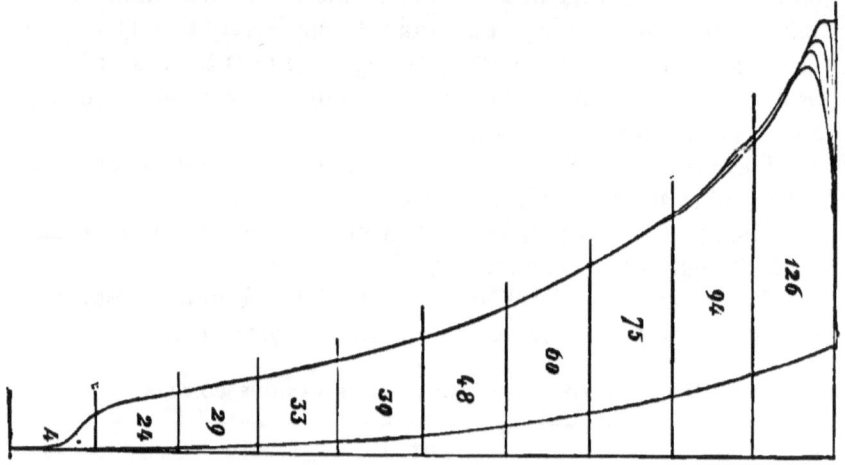

Nominal HP, 6; diam. of cylinder, 7″; length of stroke, 12″; No. of revs. 146; indicated HP, 9·05; consumpt. per IHP, 24·30 cb. ft.; consumpt. loose, 57 cb. ft.; brake HP, 7·23; consumpt. per BHP, 30·42 cb. ft.; mean pressure, 53·2 lbs.; max. pressure, 195 lbs.; press. before ignition, 48 lbs.; scale of spring, $\frac{1}{100}$″ per lb.

FIG. 60.—Diagram from Clerk Gas Engine, 6 HP.

Calculating from these diagrams the actual indicated efficiency it comes to 16 per cent. of the total heat given to the engine.

The compression space in the Clerk engines is as nearly as possible one-half of the volume swept by the piston from the exhaust port to the end of its stroke. The theoretic efficiency is therefore

$$E = 1 - \left(\frac{T'_c}{T'_o}\right)^{\gamma-1} = 1 - \left(\frac{1}{3}\right)^{·408} = 0·36.$$

The compression is higher, and therefore the theoretic efficiency of this engine is higher than the Otto, but the difficulties of proportioning the two cylinders of the Clerk engine cause a small loss of unburned gas at the exhaust ports, so that the actual efficiency is similar to that of Otto.

The mixture sent from the displacer cylinder into the motor and the space at the end of it, contains 8 vols. of air with 1 vol. of coal gas, but on passing through the upper lift valve and mixing to some extent with the exhaust there contained, i. is somewhat diluted; the heat acquired by contact with the products of combustion and with the sides of the cylinder expands the entering gases, and a temperature of not less than 100° C. is

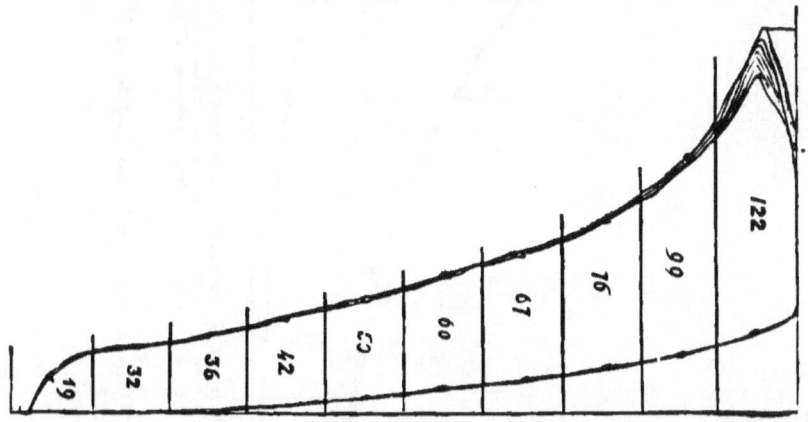

Nominal HP, 8; diam. of cylinder, 8"; length of stroke, 16"; No. of revs. 142; indicated HP, 17·38; consumpt. per IHP, 20·94 cb. ft.; consumpt. running light per hour, 70 cb. ft.; brake HP, 13·69; consumpt. per BHP, 26·58 cb. ft.; mean pressure, 60·3 lbs.; max. pressure, 195 lbs.; pressure before ignition, 49 lbs.; scale of spring, $\tfrac{1}{12}$" per lb.

FIG. 61.—Diagram from Clerk Gas Engine, 8 HP.

attained before the compression commences. The result of this is, that the displacer gases, being expanded, expel more of the exhaust gases through the discharge ports than would appear from the volume swept by the displacer piston. This volume is equal to the volume swept by the motor piston, from closing of the exhaust ports to complete in-stroke. If no expansion and no mixing occurred, the exhaust gases contained in the compression space would remain in front of the cooler explosive charge; but the heat increases the volume at least one-third, so that the volume occupied will be $1\frac{1}{3}$ times the volume swept by either piston. The volume of cylinder plus space is $1\frac{1}{2}$ vol. of cylinder,

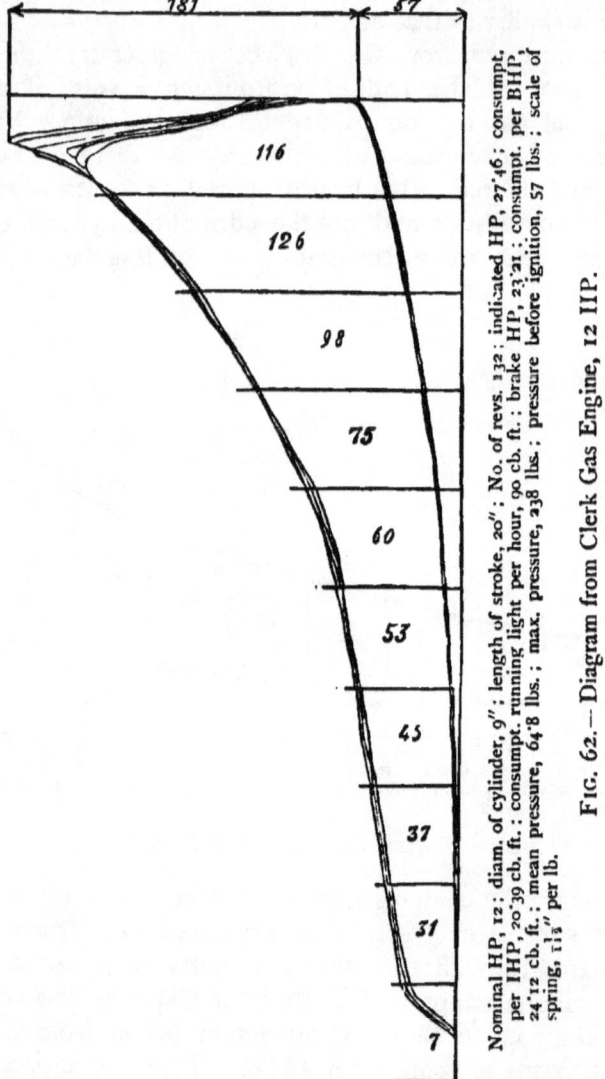

Fig. 62.—Diagram from Clerk Gas Engine, 12 IIP.

Nominal HP, 12; diam. of cylinder, 9"; length of stroke, 20"; No. of revs. 132; indicated HP, 27·46; consumpt. per IHP, 20·39 cb. ft.; consumpt. running light per hour, 90 cb. ft.; brake HP, 23·21; consumpt. per BHP, 24·12 cb. ft.; mean pressure, 64·8 lbs.; max. pressure, 238 lbs.; pressure before ignition, 57 lbs.; scale of spring, $1\frac{1}{3}$" per lb.

so that the actual exhaust gases present are $\frac{1}{6}$ vol., or $\frac{1}{9}$ of the total gases present. But mixing must occur to a considerable extent and be made very complete on the return stroke during compression. The result of all this is the production of an explosive mixture which is explosive in every part of it, and of an average composition of one volume of coal gas in ten of the mixture. The proportion of burned gases present is very slight; the only reason why any should be left is the necessity of preventing any

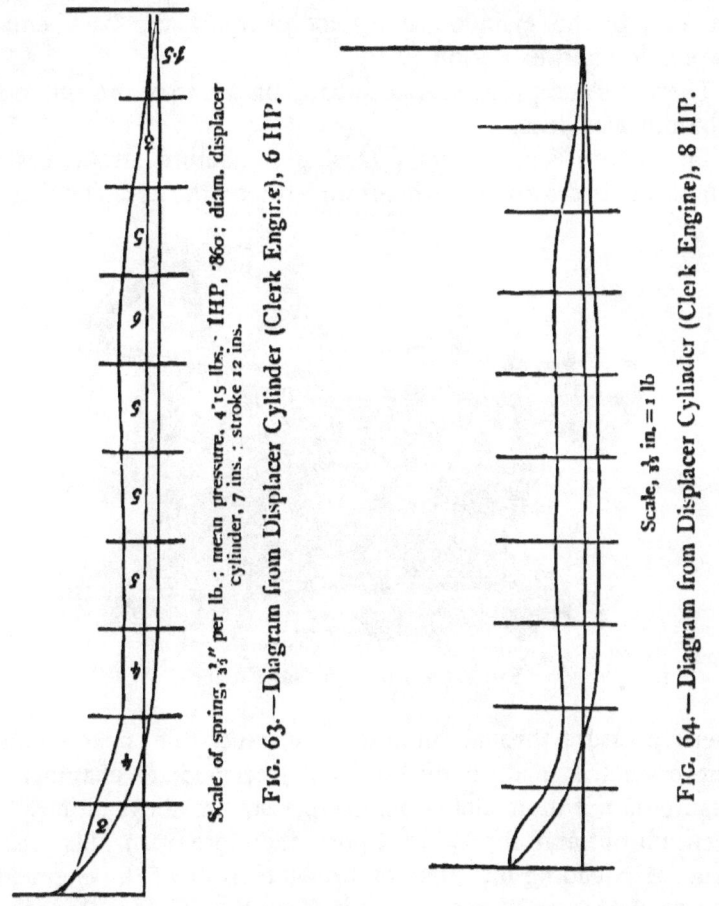

Fig. 63.—Diagram from Displacer Cylinder (Clerk Engine), 6 HP.

Fig. 64.—Diagram from Displacer Cylinder (Clerk Engine), 8 HP.

appreciable discharge of unburned gas at the exhaust ports. The mixture used is a comparatively rich one.

Tangye Engine.—Messrs. Tangye, of Birmingham, have produced an engine in which compression of the kind common to the third type is used and an ignition is obtained for every revolution when at full power. It is Robson's patent and contains only one cylinder. All the necessary operations of charging, compressing, and igniting are fulfilled with one cylinder ; it is arranged as in an ordinary steam engine. The front end of the cylinder unlike the Otto and Clerk engines is closed, the piston being provided with a piston rod, cylinder cover, and stuffing box, as in steam. The front end of the cylinder serves for charging, the back end for compression and explosion.

There is a compression space at the back end of the cylinder as in the other engines.

The action is as follows. During the return stroke, gas and air mixture is drawn into the front end of the cylinder at atmo-

FIG. 65.—Robson's Gas Engine.

spheric pressure, through an automatic valve. The next out-stroke compresses the mixture into a large intermediate chamber at a pressure of not more than five lbs. per sq. in. above atmosphere. When full out and the exhaust ports therefore open, this pressure lifts a valve leading into the compression space of the engine, discharging before it the gases contained in the cylinder through the exhaust valve and filling the cylinder and space with explosive mixture. This reduces the pressure in the intermediate reservoir to atmosphere so that the next in-movement of the piston compresses the explosive mixture upon one side of the piston and takes in fresh mixture on the other side.

When compression is completed the igniting valve acts and the explosion impels the piston; so soon as the exhaust ports open, the pressure falls to atmosphere, and then the reservoir pressure being superior to that in the cylinder, the automatic valve acts and the fresh charge enters.

Thus an explosion is obtained at every revolution by using the front end of the cylinder as displacer and storing up the pressure in an intermediate reservoir.

The governing is managed by cutting off gas supply, but is hampered considerably by the intermediate chamber. Fig. 65 is an external view of the engine, which is exceedingly neat and of substantial workmanship.

The Stockport Engine.—This engine is similar to Robson's

FIG. 66.—The Stockport Gas Engine.

in theory but the front end of the cylinder is not used for charging, the piston being made a double trunk with the crank between, and one end and one cylinder being motor, the other end and the other cylinder being displacer. Compression occurs in the motor cylinder. Fig. 66 shows the external appearance. The valve arrangements differ from those of Messrs. Tangye. It is made by Messrs. Andrew, of Stockport.

Atkinson's Differential Engine.—The description of engines of this type would be incomplete without mention of this engine, exhibited at the Inventions Exhibition for the first time in 1885. It is exceedingly ingenious and quite novel.

Fig. 67 is an elevation, fig. 68 a section, and fig. 69 a plan. The action is very clearly seen from the different positions on fig. 70.

The same cylinder serves for all purposes of the cycle; two trunk pistons, working in opposite ends of it are connected to

Fig. 67.—Atkinson's Differential Gas Engine.

the levers and from thence to the crank shaft by the connecting rods. The short rods cause the necessary actions.

In the first position, fig. 70, the pistons are at one extreme of their stroke, and are just beginning to separate. The charge of gas and air enters between them through the automatic lift valve, and in position 2, the charge has entered and the further movement of the piston is about to close the port leading to the admission and exhaust valves. The compression thus commences and in position 3 it is completed. The ignition occurs and the pistons

FIG. 68.—Atkinson's Differential Gas Engine.

now rapidly separate, the exhaust port being uncovered and the discharge commencing in position 4. By this clever method the whole operations of admission, discharge, ignition, and expansion are performed in the single cylinder with only two automatic

FIG. 69.—Atkinson's Differential Gas Engine.

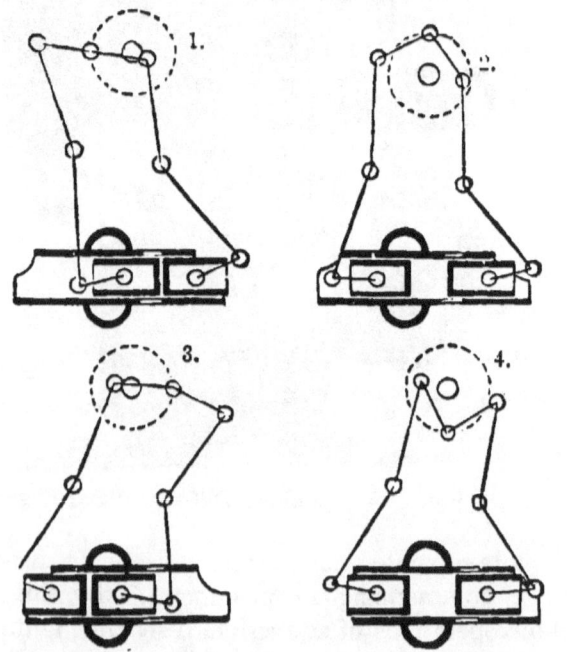

FIG. 70.—Atkinson's Differential Gas Engine.

valves which are never exposed to the pressure of explosion, the pistons acting in some part as valves and uncovering the exhaust and inlet ports when required. In the other extreme position they also act as valves, the outside piston uncovering the igniting port at the correct time. Sufficient experience has not yet been accumulated with this engine to speak positively as to its performance. To the author, the principal disadvantage appears to lie in a compression space of diameter so great, in proportion to depth, that the ratio of cooling surface to volume of hot gases is largely in excess of that common to other engines. This disadvantage will diminish the economy which the great expansion would otherwise give.

CHAPTER VIII.

IGNITING ARRANGEMENTS.

However perfect the theoretic cycle of an engine, or however admirable is its construction, in the absence of a good igniting valve the skill and energy expended is of no avail. The engine is a useless mass of metal requiring power to move itself rather than furnishing power to set other machines in motion.

In the earlier stages of gas engine manufacture, the igniting method has been the most fruitful source of annoyance and difficulty; even yet, after many years of engineering experience, the igniting valve is still the initial difficulty which the inventor must overcome before he gets the opportunity of testing his theories of heat and work in a moving machine. Quite a number of witnesses, in the shape of unworkable gas engines, in many engineers' workshops throughout Britain, attest silently but emphatically the difficulties of the igniting valve.

The problem is by no means a simple one, and the care lavished upon its solution would not be suspected on inspection of the igniting gear of any good modern engine. Much has been done, but much still remains yet to be accomplished before flame is as completely and effectively under control as steam.

In the noncompression engines the problem is comparatively simple—to inflame a volume of explosive mixture enclosed in a cylinder, so that the explosion is confined within the cylinder, and no communication is open to atmosphere. This is to be repeated regularly and with certainty at rates varying from 60 to 150 times per minute, depending upon the speed of the engine. In the earlier trials, what may be called the touch hole method naturally suggested itself; the piston after taking in its charge, crossed a small hole and sucked a flame through it into the cylinder, the hole being either small enough to occasion no substantial loss of pressure upon explosion, or covered by a small valve closing with the pressure from the interior. This is the earliest flame method. Then comes the idea of using the electric spark, and so completely closing up the cylinder, and, later on, a return to flame, using a double flame, one to ignite an intermediate one, and the intermediate flame carried in a pocket or hollow cock to the mixture. Then the idea of spongy platinum suggested by the well-known Doberiner's Hydrogen lamp. Later on the heating of metal tubes or metal masses and the ignition of the gases by contact with them. Then electrical ignition again, but this time by heating a platinum wire to incandescence. All those methods were proposed and to some extent practised long before gas engines appeared in any commercially successful form.

Ignition methods may be classed in four distinct groups.

(1) Electrical methods.
(2) Flame methods.
(3) Incandescence methods.
(4) Methods depending on 'Catalytic' or chemical action.

(1) Electrical Methods.

Spark Method.—The use of the electric form of energy seems at first sight a very convenient and easy method of getting an intense heat at any desired time and in any desired spot in the interior of a cylinder. The electric spark has long been used by chemists to explode the contents of the eudiometer in which gas analysis is effected; and the platinum wire rendered incandescent by the current from a battery has long been familiar to experimenters and is used by them for many purposes. The spark method was used in the Lenoir engine. A Bunsen's battery, a Rumkorff induction coil, and a commutator or distributor, is required in addition to the insulated points between which the spark passes in the interior of the cylinder.

Fig. 71 is drawn to show clearly the general arrangement. The Bunsen's battery A generates the current, which passes by the wires to the coil B, from which the intensity current passes to the insulated points D D by way of the distributor C. The negative pole of the coil is permanently connected to any part of the metal work of the engine; the igniting points D, D, consist of porcelain plugs seen on a larger scale at E. The porcelain is firmly cemented into the brass nut 1, and the wire 2 which passes through a hole in the plug terminates outside in the connecting screw 3, and inside is bent over the end of the plug; the other wire 4, passes through another hole in the plug, is bent over in the inside lying near the wire 2 but not touching it, it then passes through the side of the plug touching the metal of the nut. When the nut is screwed into position the one wire is in metallic connection with the cylinder of the engine, and the other is insulated from it.

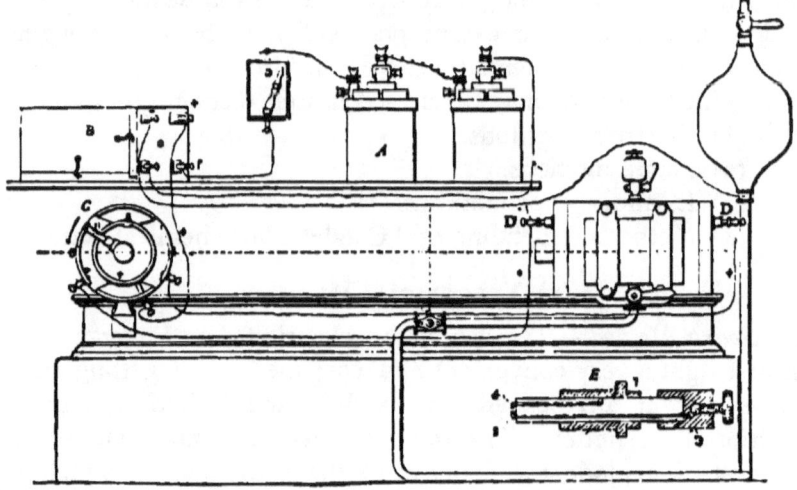

FIG. 71.—Ignition Arrangements Lenoir Engine.

The distributor C consists of an insulated metallic arm 1 rotating on the end of the crank shaft over the insulated ring 2, which is connected to the positive pole of the coil. Two insulated segments 3, 4, are connected by wires to the igniting plugs D, D; in rotating, the arm 1 comes alternately over 3, 4, and it is within sparking distance of the ring 2 as well as the segments; the sparks pass alternately to the segments and thence alternately to the opposite ends of the cylinder. The ebonite disc carrying the segments and ring is so adjusted that the spark begins to pass at either end, just as the admission valve closes. If it passed too soon the explosion would occur before the admission valve closed, and therefore would partly be lost, and at the same time would make a disagreeable noise. If it is passed too late, power is lost, because the piston is at its most rapid rate of movement and is reducing the pressure of the cylinder contents uselessly.

Notwithstanding the most careful adjustment, some time elapses between the closing of the admission valve and the explosion. When all is in good order this arrangement works very well, but should the insulation be disturbed and any short circuiting occur, the spark fails to pass between the points in the interior of the cylinder and a missed or late ignition results. This often happens in starting the engine when it is cold; the first few explosions cause a condensation of water upon the points and the spark then fails, the current passing through the water film from wire to wire without spark. The igniters then require to be uncoupled and dried. To reduce this trouble, the points are kept towards the top of the cylinder in the end covers so that any water or oil drainage may flow down and leave them dry. The difficulties of insulation, coil and battery, are so great that they did much to prevent the use of the Lenoir engine; unless the machine fell into intelligent hands it was sure to go wrong and give trouble.

The spark method has never been applied to compression engines as the compression increases all difficulties. The Lenoir igniting plug, or 'inflamer' as it was called, if put in a compression engine leaks badly and cannot be got to act efficiently. Many specifications of compression and other engines state that ignition is accomplished by the electrical spark, but the Lenoir engine alone attained any success.

Incandescent Wire Method.—This method very naturally suggests itself as a solution of the difficulties of the high tension spark; the coil is dispensed with and the current from the battery is applied directly to heat a thin platinum wire. The difficulty of insulating is very slight. The tension being low it is a matter of indifference whether the insulating material is wetted or not. The wire being constantly at a red heat cannot remain at all times in the cylinder, but is put into communication with it at proper times by means of a slide valve. Fig. 72 is a drawing of an igniting slide of this kind, as used by the author in experimental work. It acts very well indeed. The screw 1 carries the insulated rod 2, insulated by means of asbestos card-board packed into the space and screwed down firmly by the screw 3. The other wire is screwed into the metal and so is in metallic connection with the metal work of the engine. One wire from the battery connects to any portion of the engine ; the other is insulated. The platinum wire 4 is thus kept continually at a red heat, and the slide 5 moving at proper times causes the gases to be ignited to flow into the chamber containing the platinum spiral, by the hole 6, and so causes the explosion.

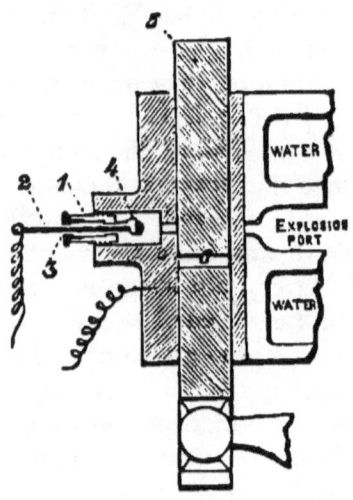

FIG. 72.—Electrical Igniting Valve (Clerk). Incandescent Platinum Wire.

There is only one precaution required in using this. The battery must not be too powerful ; if the wire is heated by it to near its fusing point, then the further heat supplied by successive explosions may cause its destruction. It requires to be kept at a good red heat and no more when open to the air : when closed up and in contact with the hot gases it will then become almost white hot ; anything above this may fuse it. The battery is of course at all times a source of care ; as it requires to be often renewed, it is only for experimental work that this arrangement answers well. In the hands of the general public it would come to grief. Hugon and many others proposed similar arrangements, but they do not appear to have worked them out.

Arc Method.—There is another electrical method. A small dynamo attached to the engine keeps up a continuous current and heavy platinum points in communication with the cylinder carry the arc. This is difficult, however, as the points constantly volatilize and require frequent renewal. This method has never come into practical use; it is described several times in specifications.

(2) FLAME METHODS.

The earliest really efficient igniting valve is that described by Barnett in his specification of 1838. It is the parent form of the most extensively used valve, the 'Otto.'

Barnett's Igniting Valve.—Fig. 73 shows a vertical section and a plan of this valve. It consists of a conical stopcock with a hollow plug; the shell contains two ports, 1 and 2—1 open to the atmosphere and 2 communicating with the cylinder. The plug of the cock has one port, 3, so arranged that it may open on the atmosphere port or the cylinder port in the shell, but cover enough being left to prevent it opening to both at the same time. In turning round it closes on the atmosphere before opening to the cylinder.

A gas jet burns at the bottom of the shell, and in the hollow of the plug, the ports 3 and 1 being long enough and wide enough to allow the air free circulation as shown by the arrows. The flame must not be too large or it will fill the whole interior with gas and prevent air getting in; the flame will then burn at the port 1 in the air and will not enter the cock. Suppose it to be burning regularly in the cock as shown in the drawing, then if the plug is suddenly turned round so that port 3 closes upon the atmosphere port 1, and opens upon the cylinder port 2, the air supply will be sufficient to keep the flame living till the mixture contained in 2 reaches it. The explosion then occurs. The port 2 is of the same shape as 1, so that the flame causes the gases to circulate the same as the air did when open to it; the mixture comes in contact with the flame by circulating through the plug. If the port 2 is made so small that no circulation occurs, then the ignition will be a very uncertain matter; as the gases will require to get at the flame by diffusion, which is a slow process, and the flame may be extinguished before they arrive at it. The explosion of course extinguishes the flame, but when the plug is again rotated to open to the air, the external flame relights it and it is ready for another ignition.

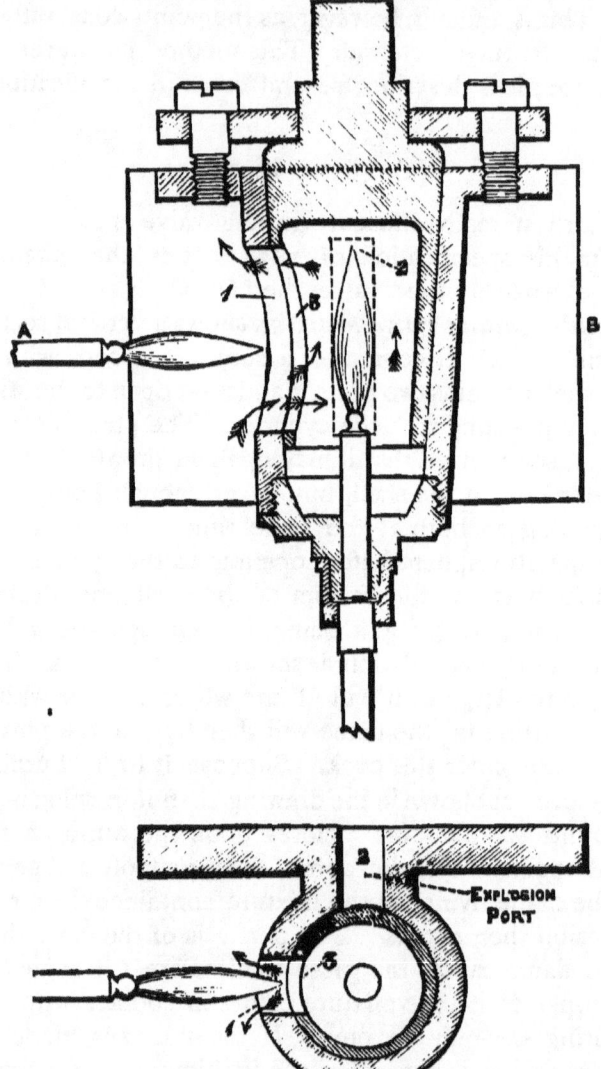

FIG. 73.—Barnett's Igniting Valve (flame).

Igniting Arrangements

Hugon's Igniting Valve.—In the small Hugon engine Barnett's method was first applied in a fairly successful manner.

The valve is shown in section at fig. 74.

The sectional plan, fig. 74, shows the internal flame lit and burning in the ignition port 1; the external flame 2 burns close to it in this position, so as to be ready to light it when wanted. The gas for the internal flame is supplied under higher pressure than that of the ordinary gas mains by a bellows pump and small reservoir through the flexible rubber pipe. For the external flame the gas is used at the ordinary pressure.

When ignition is required, the valve moves rapidly forward causing the port 1 to close to atmosphere first, and then to open to the cylinder port 3, as shown at the other end of the slide.

The explosive mixture which fills the port 3 is at once ignited and the flame finds its way from the port into the cylinder itself; the port is necessarily filled with pure explosive mixture free from any admixture with exhaust gases, as all the mixture before entering the cylinder must pass through it and so sweep before it any burned gases into the cylinder. Hence the mixture in the port will be more ignitable than that in the cylinder, as the mixture there is diluted in part with exhaust gases while that in the port is free from them.

The explosion is thus exceedingly certain and regular; when it occurs it extinguishes the internal flame and at the same time its superior pressure forces back the gas in the rubber pipe while the port 1 remains open to the cylinder.

The return of the slide again opens it to the atmosphere, and here is seen the necessity of using the gas under some pressure. Before it can relight at the external flame, the products of combustion must be expelled from the gas pipe; if the gas were under only the ordinary gas main pressure there would be no time for this, and the valve would return to ignite without a flame. The expedient of increasing pressure is somewhat clumsy but it acts fairly well. The port 1 is made large to give space for the air necessary to support the flame while the ignition port is passing from atmosphere to cylinder port. At the moment of explosion, the cylinder is completely closed from the air.

P

210   *The Gas Engine*

The explosion is therefore completely contained within the cylinder and no sound is heard.

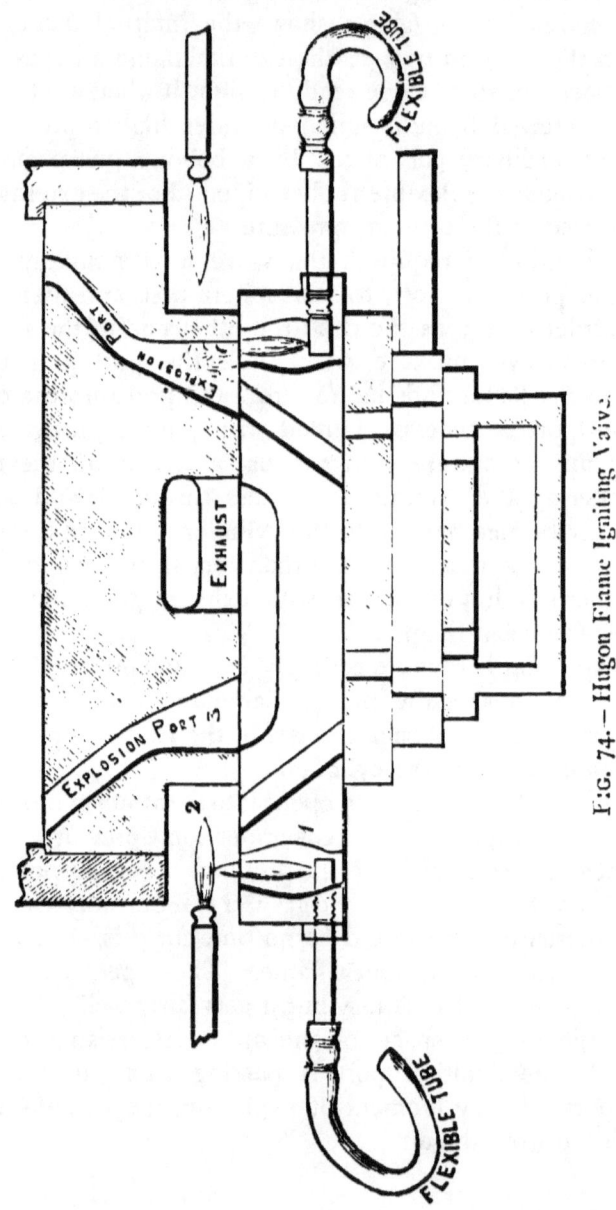

FIG. 74.—Hugon Flame Igniting Valve.

In the engine at the Patent Office Museum Mr. S. Ford considerably improved the igniting arrangement by intercepting the rush back to the gas pipe by a light check valve; he was thus able to use gas under the ordinary gas main's pressure and dispense entirely with Hugon's gas pump and reservoir. The explosion, instead of forcing a considerable volume of burned gases down the gas pipe, simply closed the check valve, which opened as soon as the igniting port reached the air again, and so gave the gas stream at once.

Otto's Igniting Valve.—The igniting valve used in the Otto and Langen engines is a further development of Barnett and Hugon's igniting devices.

As applied to the compression engine there is one alteration, very slight, but very essential.

In the Lenoir and Hugon engines, as well as the Otto and Langen, the pressure in the cylinder is the same, or in some cases less than that of the external atmosphere, that is, before ignition. It is therefore an easier matter to transfer a flame burning quietly in the air to the cylinder without danger of extinction. When the gases to which the flame is to be transferred exist at a pressure some 40 to 50 lbs. per square inch superior to that of the flame itself, it is not so easily seen how the flame is to be transferred without extinction. Generally described the arrangement is as follows. A small quantity of coal gas is introduced into the upper part of a cavity in the ignition slide; being lighter than air it remains separate from it and has no tendency to mix with the air beneath it, except by the slow process of gaseous diffusion. At the surface of contact with the air, it is ignited and burns with a blue flickering flame. The movement of the slide cuts off communication with the outer atmosphere, and very shortly thereafter opens on the admission port of the engine, but before doing this it opens on a small hole communicating with the cylinder. This hole communicates with the gas passage in the upper part of the slide, so that the gases under pressure enter and force the gas downwards, the pressure rising in the port more slowly than would occur if the main port opened at once. The pressure is therefore nearly level with that in the cylinder when the main port opens, and the flame still burning at the point or surface of junction between the gas and air, ignites the mixture. If the pressure was not raised in the igniting port by pressing the gas downwards and thereby avoiding a rush past the flame portion, the rush would often extinguish the flame and an ignition would be missed. The apparent difficulty of transferring the flame from atmosphere to

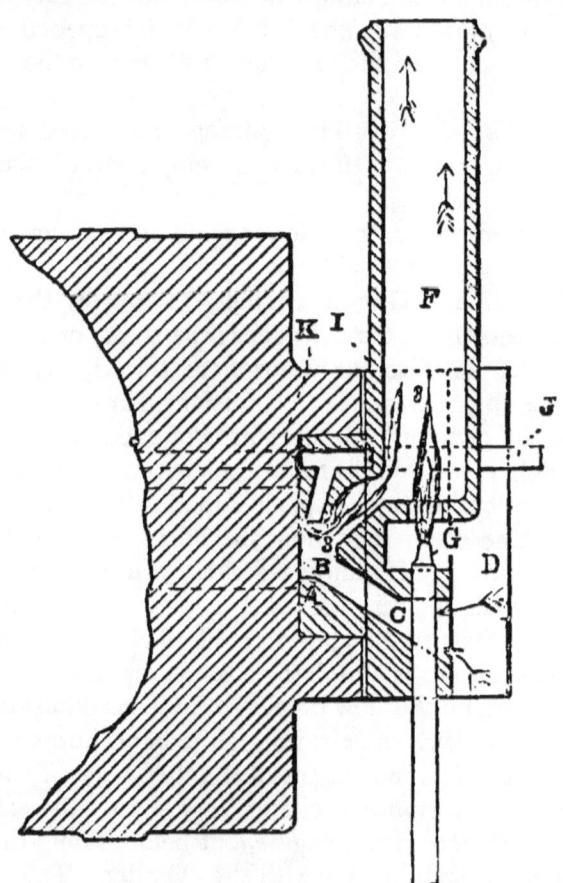

Fig. 75.—Section, Otto Igniting Valve.

40 lbs. above it is thus simply and beautifully overcome. By using a portion of gas in the upper part of the valve cavity, the difficulty of the blow back of explosion down the gas supply pipe is also overcome, as the gas supply can be cut off before the explosion or

compression pressure comes on. It is cut off just before the valve closes the flame port to atmosphere.

Fig. 75 is a vertical section showing the flame cavity in the slide, in the act of introducing coal gas at the upper part and inflaming it at the point of junction, between gas and air.

The slide A contains a forked passage B communicating at the lower passage with the air inlet C, and at the upper passage with the funnel F, which are both in the valve cover D, which holds the valve against the engine face. The jet C has a flame constantly burning into the funnel, which becomes heated, with the effect of drawing a current of air through the forked passage when its ports are in proper position; the direction of the current is shown by the arrows. The pipe J supplies coal gas which passes along the gutter I, cut in the cover and valve faces, into the forked passage C, and thence to the funnel F where it is inflamed and burns as shown. When the movement of the slide cuts off communication with the atmosphere, it also closes the gutter I and terminates the supply of coal gas from the pipe J, but the upper part of the forked passage contains gas; a flame therefore flickers as shown. Just before B opens on the port L, fig. 76, the hole K, fig. 75, opens and the pressure from the explosion space causes a flow into B, forcing before it the gas contained in the hole, thereby intensifying the flame by making the gas pass more into the air and bringing about the equilibrium of the pressures. When B opens on L, the flame is a vigorous one, and at once fires the whole charge in the explosion chamber. Fig. 76 shows the slide with the port B at the moment of opening on L. Fig. 77 is an end elevation of the valve and cover, showing the ports and gutters dotted and lettered, position same as in fig. 76. The method is carried out completely and is a very perfect one indeed; it is somewhat slow in action, depending as it does on a proper ventilation of the forked passage and the complete replacement of the burned products by fresh air before the gas can burn properly in the cavity. If the engine be run more rapidly than the draught of the funnel can clear out the passage from the burned gases, then the flame cannot be lit in it and an ignition will be missed.

It is a method exceedingly successful when ignition is not required too frequently, but very troublesome and uncertain for rapid ignition. The Otto and Langen engine only made 30 ignitions per minute, and the Otto compression engine makes but 80

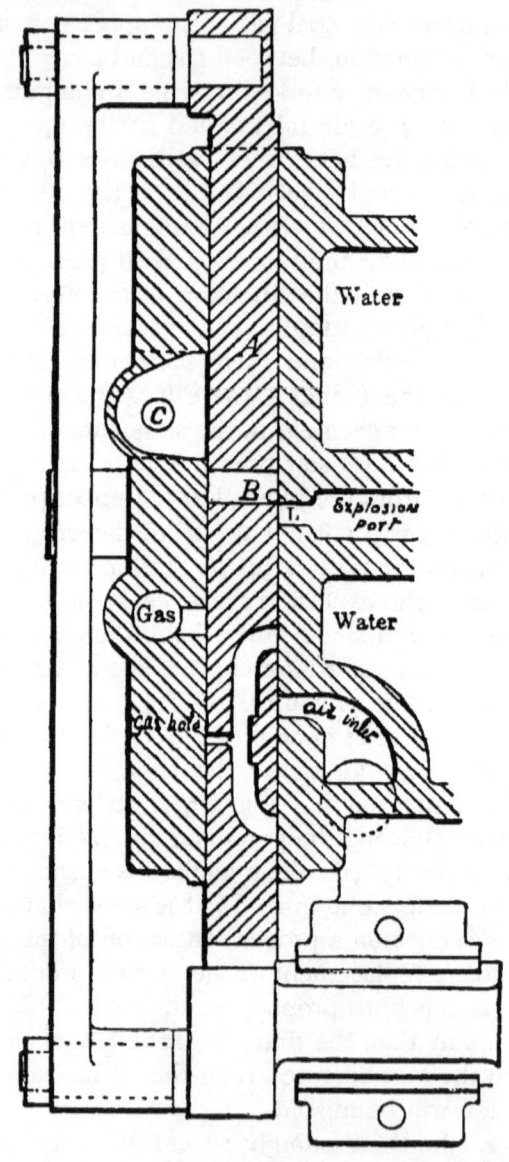

FIG. 76.—Sectional Plan, Otto Igniting Valve.

ignitions per minute at full power; its efficiency is good at these rates, but at 150 per minute it is too slow in action.

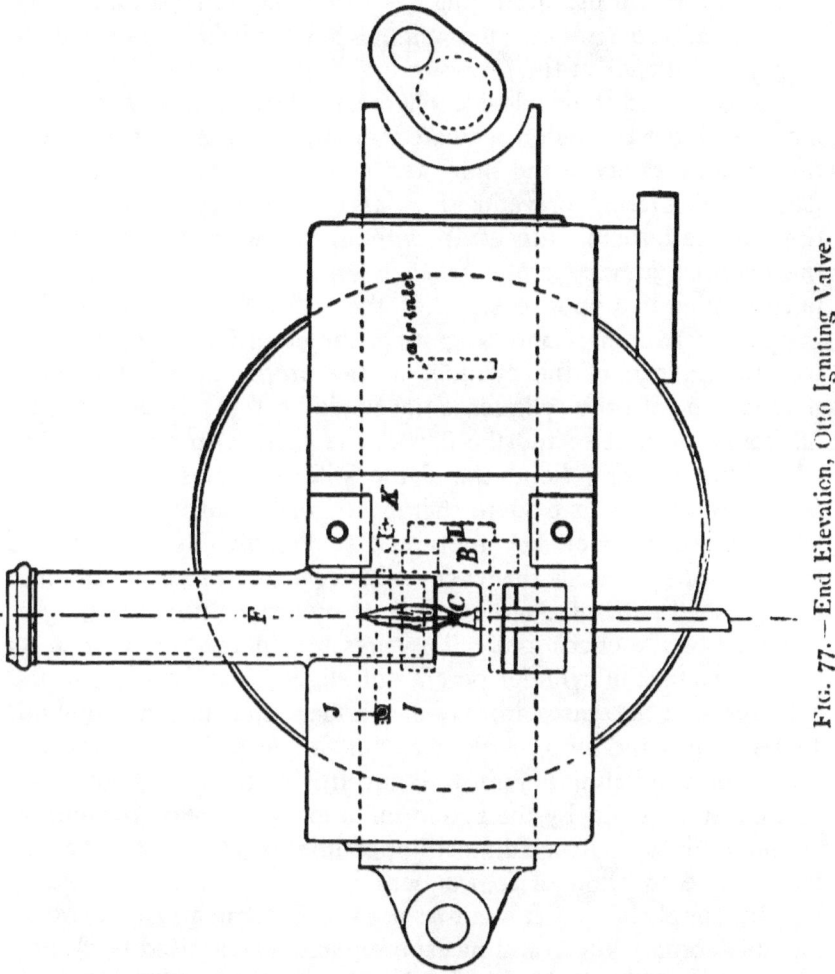

FIG. 77.—End Elevation, Otto Igniting Valve.

Clerk's Igniting Valve.—The method of igniting the charge used by Clerk is quite different from the other flame methods already described; the difference is necessitated by the greater rapidity of ignition in engines with an impulse for every revolution.

To ventilate the igniting port in the Otto and Hugon slides requires time, which cannot be given when the frequency of the ignition approaches 150 to 200 per minute.

To meet this difficulty the author has invented several methods both flame and incandescence, but the one to be described is that at present in use in his engines ; it is very reliable and rapid, as many as 300 ignitions per minute having been made with it experimentally, or at the rate of 5 ignitions per second.

A portion of the explosive charge is allowed to pass from the motor cylinder through a regulated passage to a grating placed at the end of a cavity in the slide, and is there ignited by a Bunsen flame ; the grating prevents the passage back of the flame, and the mixture burns in the cavity without requiring the presence of the external atmosphere. At each end of the cavity there is a port opening to opposite sides of the valve, the one for lighting the gases streaming from the grating, the other for communicating with the interior of the cylinder at the proper time. The communication with the cylinder is not made until the outer port cuts off from atmosphere, and the flow of the gases is so regulated that while this is being done, the flame still continues to be fed by fresh supplies. It is evident that if too great a current be sent in, the pressure will soon become equal to that in the cylinder, and then the flow towards the cavity will cease and the flame become extinguished ; this is guarded against by proper proportioning of the blow by the check pin. The pressure in the cavity when its port opens on the cylinder port is still slightly less than that in the cylinder, and the gases from the cylinder enter and are ignited. By using gas and air already mixed in proper proportion, the necessity of ventilating is removed, and it is made possible to ignite at the rate required by the system of impulse at every revolution. Without this it would be almost impossible to get a passage cleared out in time to allow of so frequent ignition, by a coal gas flame burning simply in air. It was first used by Clerk in an engine working in February 1878, and has subsequently been used by Wittig and Hees and by Robinson in the Tangye engine. In the form here described it was first used by Clerk in November 1880.

Fig. 78 is a sectional plan of the igniting slide and cover as well as the passage into the combustion space. The valve 1 contains the cavity 2, furnished at the ends with the ports 3 and 4 ; at the end 3 is placed the grating 5, communicating behind with the explosion port 6, by a small hole 7 and a gutter in the valve face, showing at fig. 78. A long pin 8 screwed into the end of the slide controls the gases entering the space behind the grating, and if need be can cut off communication altogether. When the valve is in the position shown in the drawing, the mix-

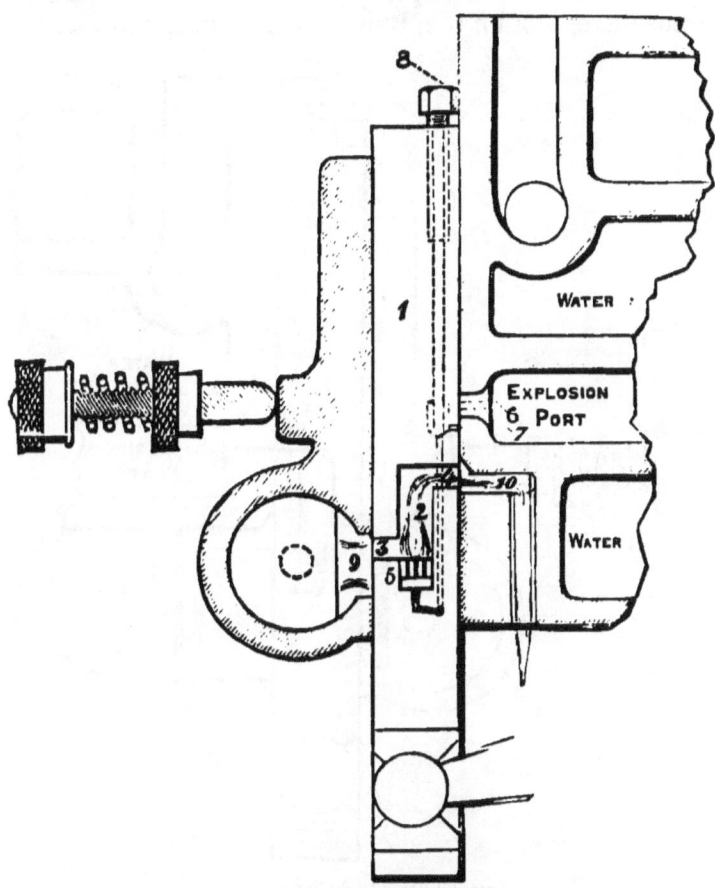

Valve in position of flame lighting at external flame.

FIG. 78.—Sectional Plan, Clerk Igniting Valve.

ture is beginning to flow through the grating into the space 2, and is ignited by the Bunsen flame 9 lying up against the valve face. The Bunsen flame lies so close to the grating that immediately inflammable mixture comes, it is lighted before it can get time to fill the cavity; if allowed to accumulate in the cavity before lighting, a slight explosion ensues and a disagreeable report is produced. The flame at the grating burns in the cavity, discharging into the passage 10, and from thence to the atmosphere. The movement of the slide cuts off communication with the atmosphere, first on the Bunsen flame side, and then on the

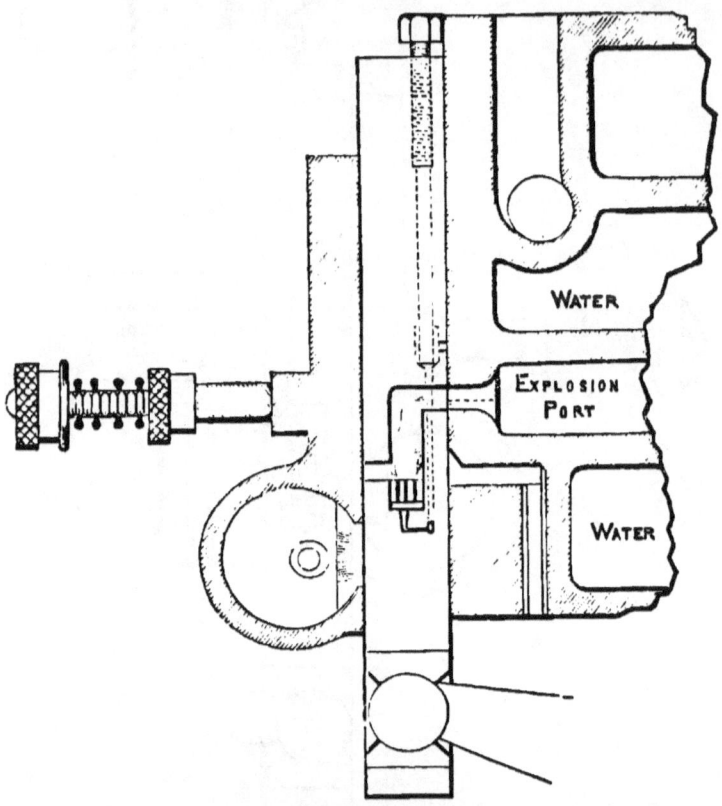

Internal flame exploding mixture.
FIG. 79.— Sectional Plan, Clerk Igniting Valve.

inside of the valve; very shortly after, the port 4 opens on the port 6 leading to the cylinder, and the gases then taking fire communicate the flame to the whole contents of the compression space. In fig. 79 the flame port in the valve is full open on the explosion port of the engine. The slide then moves past the port and back to the first position, where the operations described are repeated and igniting again occurs.

This arrangement is very rapid in action, and is capable of igniting with the utmost regularity at a rate so high as 300 times per minute, which is far in excess of the requirements of the engine. Fig. 80 shows the Bunsen flame burning against the face of the valve, ready to ignite the gaseous mixture.

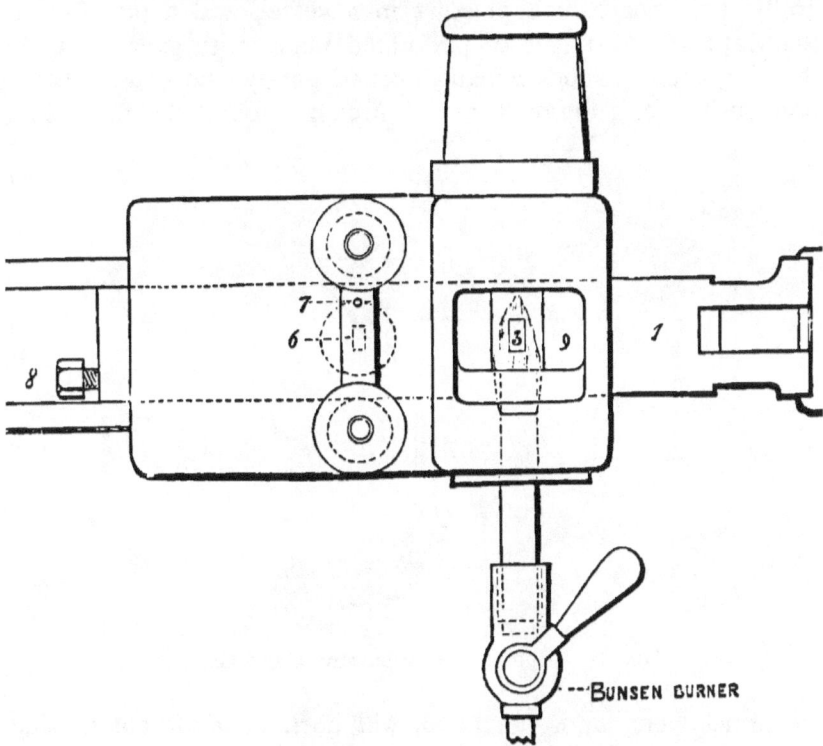

FIG. 80.—End Elevation, Clerk Igniting Valve.

Brayton's Flame Ignition.—The Brayton method of ignition has already been described shortly in the description of the engine. It is so beautiful and instructive that it merits further discussion.

The action will be made clearer by describing a well-known laboratory experiment (fig. 81).

A piece of wire gauze, *a*, held a few inches from the Bunsen lamp, *b*, the gas being turned on, will prevent the flame when lit above it from passing back through the gauze to the burner. The gauze may be moved through a considerable distance from the Bunsen tube without extinguishing the flame. The mixture of gas and air streaming from the Bunsen passes through the gauze, and, although igniting above, the heat is so rapidly conducted away by the gauze, that the flame cannot pass through its interstices back to the lower side. If an explosive mixture be confined under say 30 lbs. per square inch pressure in a vessel, and a pipe from it (fig. 82) leads to a pair of perforated plates with gauze between them, a, then the cock b being opened gently (the valve c being previously open), the mixture will stream through the plates into

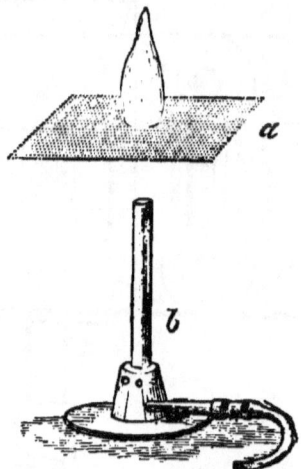

FIG. 81.—Bunsen Flame burning above Gauze.

the atmosphere, and, if ignited, will burn at a without passing back. If the cock b is opened suddenly a greater rush of flame will occur, diminishing again if it is partly closed.

So long as enough mixture passes to preserve alive the flame at a, then any increased quantity passing from the reservoir will be burned; the little flame increasing or diminishing as the opening of the stop-cock valve is increased or diminished.

The action of the ignition in the Brayton engine is exactly similar. The pressure on the flame side of the grating is slightly below that existing on the other side; the stream of cold gases entering the engine cylinder immediately becomes flame on the grating, and so expands, the volume of flame being changed as required by the valve action of the engine.

This method is most successfully carried out in the Brayton engine. The lack of economy is not due to the ignition, but to the use of it under unsuitable circumstances. Without doubt this

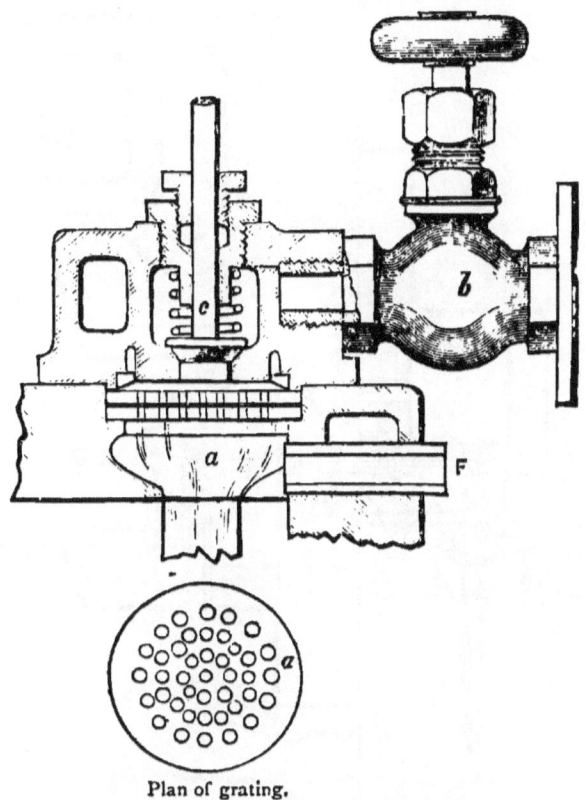

Plan of grating.

FIG. 82.—Brayton Grating and Valve.

system, in a better combination, will come largely into use in future and larger gas engines. It is unsuited for cold cylinder explosion engines, but admirably adapted for hot cylinder combustion engines of the second type.

(3) Incandescence Methods.

The ignition of explosive mixtures by contact with heated metallic surfaces has often been proposed, first by the late Sir C. W. Siemens, and after him by the American, Drake. Dr. Siemens, in one of his gas engine patents, proposes to ignite the

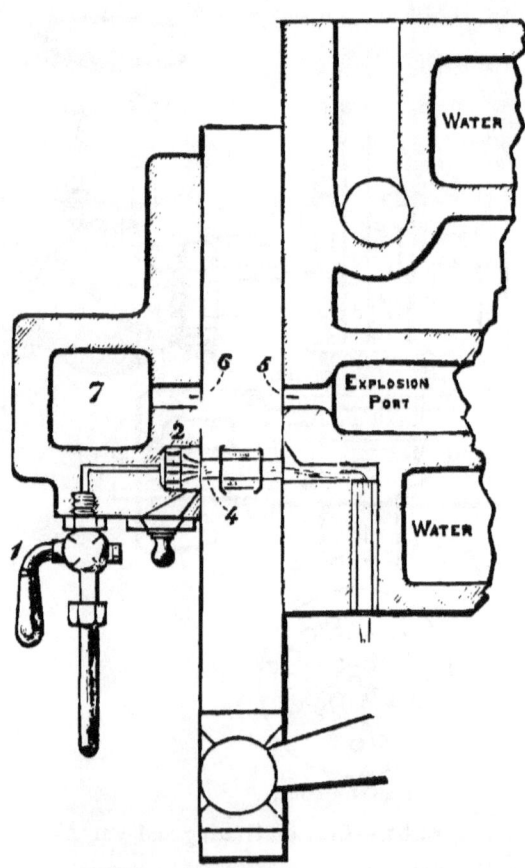

Fig. 83.—Sectional Plan, Clerk Incandescent Platinum Igniting Valve.

mixture by passing it through an iron tube, which is heated to redness by a flame outside of it.

Drake constructed an engine in which the ignition was effected in a similar manner. The difficulty is found in the rapid oxidation of the tube, and the consequent necessity for frequent renewal. Frequent attempts have also been made to heat a portion of the interior surface of the cylinder, so that at a suitable time the mixture might be exposed to it and fired.

The first arrangement of incandescent ignition successfully applied to a compression engine is the invention of the author, and is described in his patent, No. 3045, 1878. It was used in an engine exhibited at the Royal Agricultural Society's Show, Kilburn, in 1879 (July).

Clerk's Igniting Valve.—Fig. 83 is a sectional plan of this valve in position. Fig. 84 is a separate view of the valve looking upon the face, and fig. 85 is the platinum cage, full size, taken out of the valve.

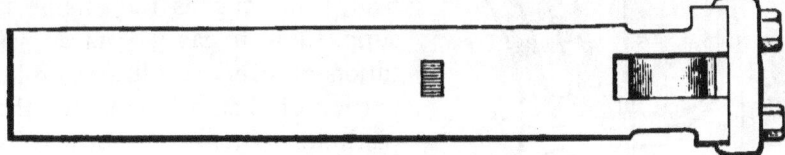

FIG. 84.—Face of Valve with Platinum Cage.

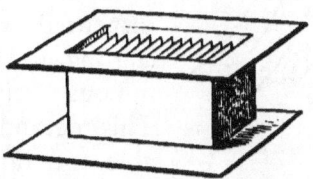

FIG. 85.—Platinum Cage.

The platinum cage consists of a box of platinum plate, with numerous platinum ribs running across it. They are secured by rivets running completely through, small platinum washers serving to keep the plates at equal distances. The valve receives this cage in a cavity, and it is tightly packed in its place with asbestos and slate packing, a covering plate screwed down upon it securing the whole in position. To start the engine, the reservoir containing gas and air under pressure is opened; the small tap, 1, then opened allows mixture to flow through the diaphragm 2 (made like the Brayton grating), and the mixture is ignited at the small door 3, which is then closed. The flame flows through the platinum cage, heating up its plates to a white heat in a few seconds. On opening the starting cock of the engine, it moves, and brings the igniting port 4, on the cylinder port 5, at the same time opening on the port 6, in the cover, leading into the cavity 7. The mixture in the cylinder then rushes through the cage, becoming ignited, and the explosion reaches the cylinder; the cavity 7 is so proportioned that each ignition sends a measured quantity of flame through the cage into it; the heat of the explosion at every turn therefore supplies heat to the platinum. This added heat is sufficient to keep it at a white heat. So long as the engine is supplied with gas it gets an ignition at every revolution, and a portion of that heat goes to the platinum to make up for loss by conduction. The heating flame used in starting the engine is dispensed with immediately on starting, and the engine runs continuously without outside flame. This method is exceedingly reliable and rapid, but is not suited for the governing arrangements of small engines.

Siemens' Tube Method.—Fig. 86 is an arrangement of Siemens' method, as used by Mr. Atkinson in his 'Differential' engine, exhibited at the Inventions Exhibition. The wrought iron tube 1 is heated by the Bunsen flame 2, the non-conducting casing 3 preventing loss of heat; the piston at the proper time uncovers the hole 4 into which the tube is screwed, and the mixture entering under pressure becomes ignited. In other engines the tube is

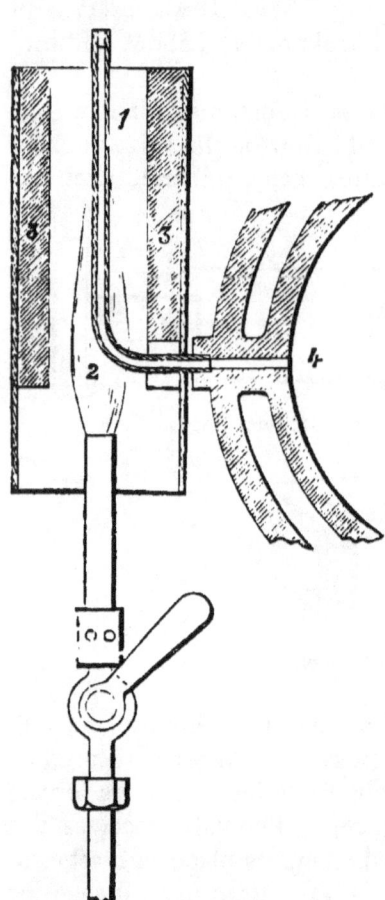

FIG. 86.—Hot Tube Igniter.

caused to communicate with the cylinder by a valve. This modification is exceedingly simple and works well; care must be taken to avoid overheating, or the explosion may rupture the tube. It is inexpensive and easily renewed when disabled by oxidation.

(4) Methods depending on Catalytic and Chemical Action.

The well-known property of spongy platinum of causing the spontaneous ignition of a stream of hydrogen or coal gas directed on it in air, has been proposed as a means of ignition by Barnett (1838). In the arrangement he describes, the platinum is contained in a little cup screwed into the cylinder cover, and the compression of the mixture causes its ignition by contact.

Platinum, however, soon loses this property, and the action is at best too slow for use.

All flame methods of course depend on chemical action, but one proposal has been made, to use the property possessed by phosphorated hydrogen of igniting spontaneously in contact with air. The phosphorated hydrogen is conducted in small quantity into the mixture to be exploded at every revolution, and its combustion causes ignition.

This proposal has never been carried out in practice.

Summary.—To the author's knowledge no other systems of ignition have been proposed; the flame methods are best suited for small gas engines and will probably continue in use. Considerable improvements may still be effected in ignition valves, and it is possible that external flames may be entirely done away with in future engines. It is somewhat humiliating to the inventor to watch a powerful gas engine at work, developing say 30 horses, and to know that he can at once change the whole and make the engine powerless by blowing out the external flame.

A combination of flame and incandescence methods will doubtless overcome this difficulty, and make the gas engine act without visible flame and without the danger of extinction from draught, to which the present igniting flames are subject.

It is improbable that either the first or fourth methods will again find favour, the electric methods give too much trouble and are at best uncertain.

CHAPTER IX.

ON SOME OTHER MECHANICAL DETAILS.

A GOOD working cycle and good igniting arrangements are the two most important factors in the successful working of a gas engine, but there are other matters whose importance is only secondary to those. The governing gear, the oiling gear, and the starting gear, are of the greatest importance.

These matters will now be described.

The Governing Gear.—In the earlier gas engines, including Lenoir and Hugon, the governing was attempted precisely as is done in the steam engine, the source of power being regulated by throttling. A centrifugal governor acted upon a throttle valve regulating the gas supply, diminishing it when the speed became too great and increasing it when the speed fell.

This was a very bad and wasteful method, as the engineer will at once recognise from his knowledge of the properties of explosive mixtures.

The limits of change allowable in the proportions of gaseous explosive mixtures are very narrow, the gas present ranging from $\frac{1}{7}$ to $\frac{1}{18}$ of the total volume. A mixture containing $\frac{1}{7}$ of its volume of coal gas in air has just sufficient oxygen to burn it and no more; any further increase of gas will pass away unburned, there being insufficient oxygen present for its combustion.

This is therefore the richest mixture which can be used with any economy.

A mixture of air and gas containing $\frac{1}{18}$ of its volume of gas is in the critical proportion; any further dilution, however slight, will cause it to lose inflammability altogether. The governor may act in changing the proportion of gas and air between those limits, that is, the explosion may be so reduced by dilution that it gives only half the power per impulse obtainable with the strongest mixture.

Any further dilution causes the engine to miss ignition altogether, and discharge the gas it has taken into the exhaust pipe, without obtaining any power from it. If, therefore, the governor acts by throttling, the valve is only closed enough to cause the mixture to be so weak as to miss fire; as soon as that point is reached the valve will be closed no further, because at that point the speed of the engine will cease to increase. Fig. 7, p. 14, shows the governor in action upon a Lenoir engine.

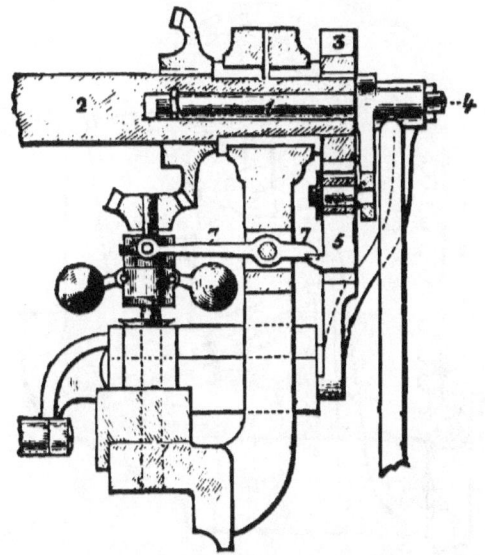

FIG. 87.—Section showing Otto and Langen Governor.

In modern compression engines the great loss of gas occasioned by throttling is avoided by never diluting the mixture. Instead of keeping up the same frequency of impulses but of less power, as done in the steam engine, the gas is either full on or full off, that is, the governing is effected by diminishing the frequency of the impulses instead of diminishing their power.

In the specification of the Otto engine, 1876 (2081) the governing is described as being effected by reducing the power of the explosion. This is more impracticable in the Otto engine than in the Lenoir, because, owing to the dilution of the charge by the exhaust gases or air, the range of change in mixture is smaller. The strongest mixture does not exceed 1 of coal gas in 8 of other gases.

Governing—Otto and Langen Engine.—In this engine, in its latest and best form, the governing is effected by missing impulses. When the engine has received an impulse, the increase in speed causes the governor to move a lever which disengages a pawl from a ratchet, and so prevents the piston being raised and the charge drawn into the cylinder. When the speed has fallen sufficiently

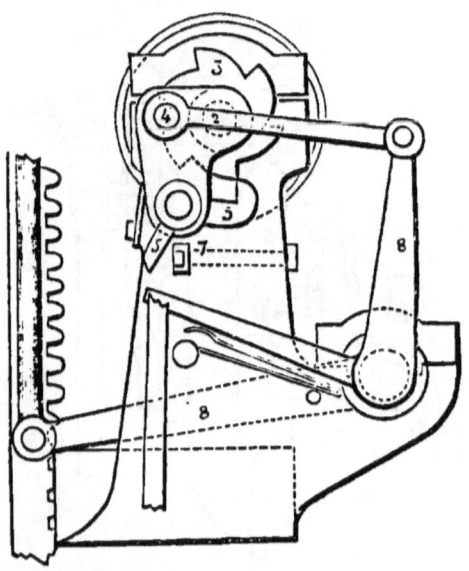

FIG. 88.—Otto and Langen Governor, showing Pawl and Ratchet.

the lever liberates the pawl, and the piston is then raised, taking in the charge and exploding it. Fig. 87 is a sectional elevation of the governing arrangements. The auxiliary shaft 1 is driven from the main shaft 2 by the clutch 3, but the crank 4 and shaft 1 receive motion from 2 only by means of the pawl 5 falling into the ratchet 3; so long as the governor lever 7 remains in the position shown, the pawl is kept from engaging and the piston and valve remain at rest; so soon as the governor lever 7 liberates the pawl, then it falls into the ratchet wheel by a spring and the auxiliary shaft receives one turn ; the crank 4, connected to the lever 8 (fig. 88), lifts the rack, and the piston takes in its charge ; at the same time the valve opens to gas and air, then, when the piston is full up, brings on the igniting flame. The explosion occurs and shoots up the piston, which on its down stroke accelerates the motion of the power shaft, and if the limit of speed is exceeded, the governor lever again interposes and prevents the charge and explosion till the speed falls.

When running without any load, the two horse engine tested at Manchester by Clerk required only 6 ignitions per minute, consuming, including side lights, about 25 cubic feet per hour. The shaft therefore ran as many as 15 revolutions merely by the power stored in the fly-wheels.

The governing is effective but irregular.

Governing—Otto Engine.—The speed of the Otto compression engine is governed by diminishing the number of impulses given to the crank ; whenever the normal rate is exceeded, the governor so acts that the gas supply is completely cut off for one or more strokes of the engine, no impulse being given till it falls again.

One arrangement very commonly in use is shown at fig. 89.

The cam 1 upon the auxiliary shaft 2 is arranged to strike the wheel 3 upon the lever 4, opening the gas valve 5 at the beginning of the stroke and keeping it open till the end of the stroke of the piston; the gas passes from the gas valve by a passage to the holes in the slide, when it streams into the air current entering the engine by the admission port. Whenever the speed becomes high enough, the governor 6 by the lever 7 shifts the position of the cam 1 upon its shaft, so that the wheel 3 does not strike it; the gas valve 5 therefore remains shut for that stroke, and the piston draws air alone into the cylinder. When the piston returns and compresses the charge, the igniting flame enters as usual, but there being no explosive mixture there, the piston moves out again without impulse, expanding and discharging, charging and compressing an uninflammable charge, till the reduction of speed calls again for an impulse ; the first ignition after the engine has made several revolutions without gas is always more powerful than the normal one, because no exhaust gases being there the charge mixes in the space with pure air and is not heated previous to explosion.

230   *The Gas Engine*

The arrangement in different Otto engines varies from this, but the principle is always the same.

Fig. 90 is a recent and very clever governing arrangement as used in the smaller Otto engines.

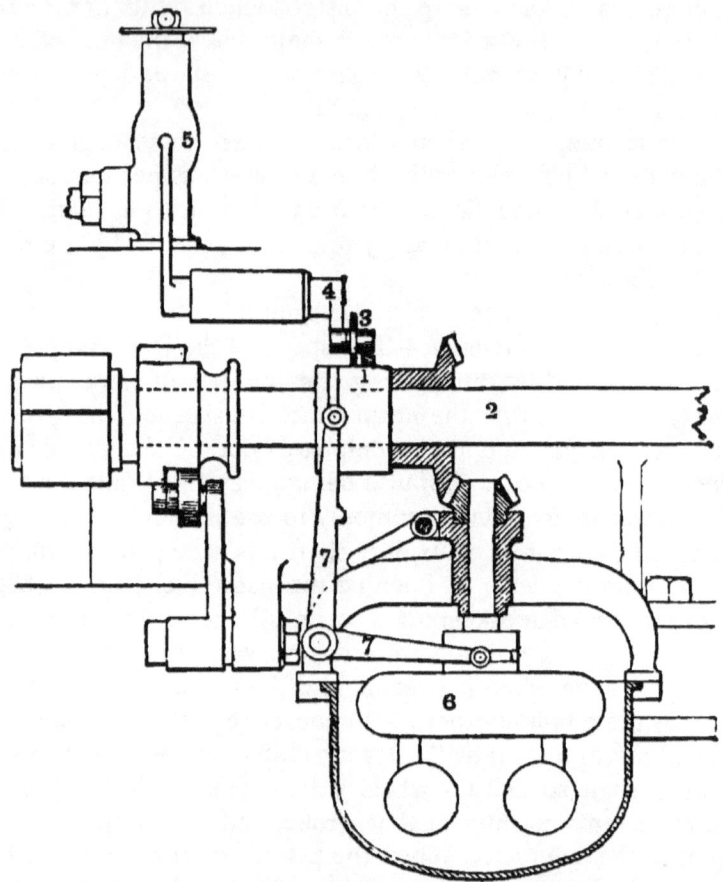

FIG. 89.—Otto Governor and Connecting Gear.

The ordinary governor is entirely dispensed with, and the valve itself carries a pendulum which governs.

The pendulum 1, hanging from the pin 2 in the slide valve 3, carries the long steel blade 4, which usually strikes the stem 5, and opens the gas valve at the same time as the slide opens to the air. Whenever the speed is exceeded, however, the motion of the valve in the direction of the arrow, exceeding a certain rate, the pendulum 1 is left behind and depresses the steel blade 4, which therefore misses the gas valve stem and for that revolution no gas enters. So long as the speed is sufficient to swing back the pendulum no gas enters; as soon as it is insufficient to cause the pendulum to leave its resting position against the valve, then gas is admitted.

As the pressure of the edge of the steel plate upon the valve stem is in direct line with the centre of the pin upon which the pendulum hangs, there is no tendency to move it, that is, the governor does not furnish the power to open the gas valve. In all the Otto governing arrangements this principle is adhered to; the

FIG. 90.—Otto Pendulum Governor.

governor never furnishes the power to move the gas valve, but only signals to the engine the proper time to give the motion, the motion being always taken from the engine itself.

In electric light engines, which must give the impulse for every two revolutions with some change of power, the gear is modified; instead of complete cut-off as first described, the cam upon the shaft is made in several steps, so that the wheel upon the gas lever is shifted from one to another as shown in fig. 91, where 1 is the gas cam, and 2 is the wheel upon the gas lever. Those steps are made to diminish supply of gas as much as possible without missing ignition, so that within narrow limits of changing load, the engine may retain its frequency of impulse. Whenever this range of permissible variation is exceeded, the wheel slips entirely off the cam, and the engine then governs in the ordinary manner.

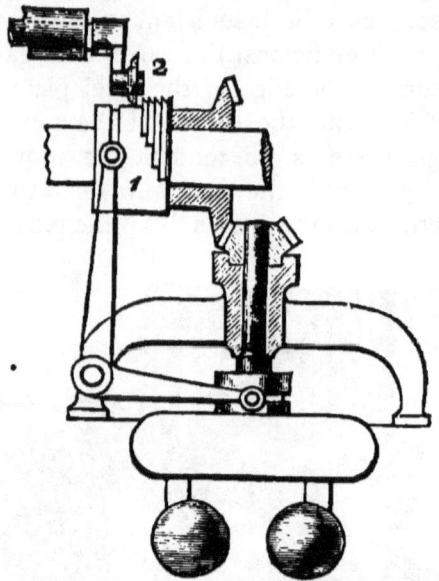

FIG. 91.—Otto Electric Light Governor.

Governing—Brayton and Simon.—In the Brayton gas engine the governing was effected precisely as in the best steam engines, by varying the point of cut-off. The entering flame was cut off, sooner or later, as determined by the governor of the engine; and the admission of gas and air to the pump was simultaneously regulated, the amount entering being diminished to keep the pressure in the reservoir constant. The diagram, fig. 45, p. 158, shows that the variable cut-off acted well.

Fig. 92 shows the governor of the petroleum engine.

The cam 1, which opens the admission air valve on the motor cylinder, is made tapering, so that the point of cut-off becomes earlier and earlier as it slides in the direction of the arrow.

The supply of air was thus diminished. In this engine the supply of petroleum could only be diminished by hand, two screws on the oil pump, when screwed upwards, altering the connecting rod between the plunger and the eccentric, giving more or less free movement, and thereby diminishing the throw of the pump.

The air supply to the engine was not diminished, so that the pressure in the reservoir increased, and was blown off at a safety valve placed upon the engine. This was a wasteful method.

The regularity of this engine in running was very great, being far superior to any of the modern compression engines. It was, however, not at all economical.

Simon's engine presented no new feature in its governing arrangements. They were quite similar to Brayton.

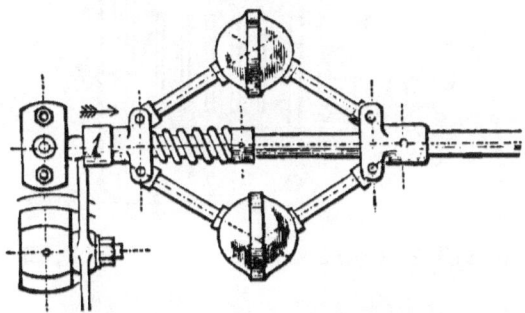

FIG. 92.—Brayton Governor.

Governing—Clerk Engine.—The governing gear now used upon this engine is the design of Mr. G. H. Garrett, Messrs. L. Sterne & Co.'s works' manager. It is shown at figs. 93 and 94.

It consists of a gridiron slide placed between the upper and lower lift valves. So long as the engine is at full power, the slide 1, fig. 93, is moved by the lever 2, fig. 94, from the ignition slide of the engine already described, and remains open during the forward stroke of the displacer piston.

The charge of gas and air therefore enters during the whole stroke, and is sent into the motor cylinder to be compressed and ignited at the proper moment. If, however, the load is lessened, and the speed increases, and the governor 3 acts, it moves the lever 4, which then catches the lever 2, and prevents the spring 5 from taking the slide 1 back and opening it. The displacer then discharges its contents into the motor cylinder, but on its next out-stroke, the valve 1 being closed, it gets no charge but the

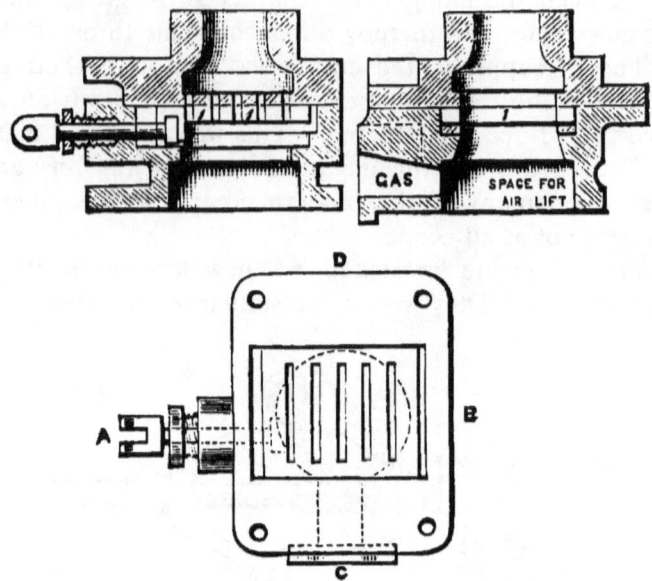

FIG. 93.—Sections and Plan, Governor Slide, Clerk Engine.

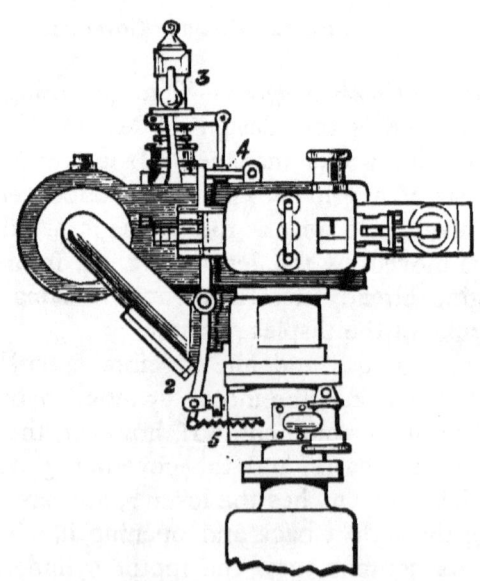

FIG. 94.—Clerk Engine showing Garrett Governor Gear.

piston moves out, forming a partial vacuum behind it. The motor cylinder, therefore, receives no charge from the displacer cylinder, and the motor piston compresses and expands alternately the burned gases behind it, while the displacer piston moves out and in, expanding and compressing likewise. This goes on till the governor signals reduction of speed, and disengages the lever 2, by pushing down the lever 4, so that the spring 5 opens the slide 1, and the engine gets a charge.

This method works very well and economically; it is necessitated by the clearance space unavoidable in the Clerk engine between the motor and displacer cylinders. If gas were cut off as in the Otto, that space filled with mixture would be lost every time the governor acted.

Governing—Tangye Engine.—Messrs. Tangye's gas engine is now controlled by a very ingenious governor, the invention of

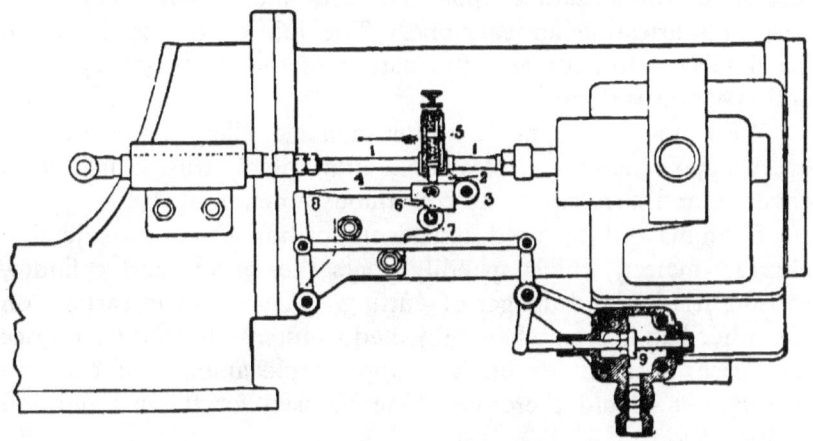

FIG. 95.—Governing—Tangye Engine.

Mr. C. W. Pinkney. It is shown at fig. 95. The rod 1 1, moved to and fro by an eccentric, carries with it the bracket 2, into which is fixed the pin 3; on this pin the lever 4 is swung, and moves to and fro with the bracket; the lever is pressed gently downwards by the spring 5, and the lower part of the lever is formed into an incline at 6, so that as it moves the spring presses it against the roller 7. So long as the engine does not exceed its proper speed, the lever 4 does not rise above the position shown in the figure when it is moving in the direction of the arrow, and accordingly its knife edge end strikes the lever 8 and, acting through the intermediate links, opens the gas valve 9. The engine gets its charge of gas every time the gas valve opens. If the speed becomes too great, then the upward velocity given to the lever 4 by the stationary roller 7 forcing against the incline is such that the knife edge lever 4 rises above the end of the lever 8, and the gas valve remains closed. When the speed falls sufficiently the lever 4 again strikes the lever 8 and opens the gas valve.

The incline governor works well and is exceedingly sensitive to change in speed: by altering the compression upon the spring 5 the speed of the engine can be varied.

Oiling Gear.—In the steam engine the comparatively low temperature of the steam within the working cylinder and the fact of its condensation upon the walls and piston renders the task of lubricating an easy one. The lubrication need not be absolutely continuous and the nature of the oil may vary much and no harm is done.

With the gas engine, the intense flame filling the cylinder at every stroke quickly destroys the film of oil with which it is covered, and necessitates its continuous renewal.

If animal oil be used, its decomposition leaves considerable charred matter, which speedily coats the piston and cylinder, causing friction and danger of cutting. A good hydrocarbon, on the other hand, even when subjected to intense heat, decomposes into gases without leaving any appreciable amount of carbon: mineral oils should therefore alone be used for the cylinder and ignition slide.

The amount of oil required for these parts is small per day, but it must be regularly applied; the burned film removed from the surface of the cylinder at every explosion must be regularly replaced or abrasion of the surfaces would speedily ensue.

In the Otto engine the oil required is supplied during the whole action of the engine; it commences with the movement of the engine, continues so long as it is running, and stops when motion ceases.

Fig. 96 shows the Otto oiling cup, one of which is placed, as shown in the drawing, fig. 97, at the middle of the cylinder to lubricate the piston and slide; the pipes 4 and 5 lead to the piston and slide.

The pulley 1 is driven slowly from the auxiliary shaft by a strap, and as it rotates it carries the wire 2 round on the pin 3,

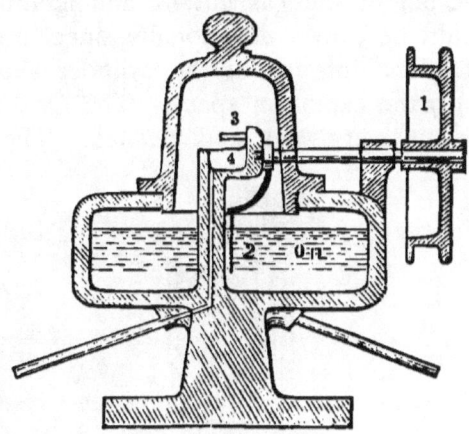

FIG. 96.—Illustration of action of Otto Oiler.

fig. 96, alternately dipping into the oil and wiping it off to the pin, from whence it drops into the trough 4 and runs by a hole into the tubes. The amount of oil so discharged can be regulated by the diameter of the wire. The oil flows along the pipes 4 and

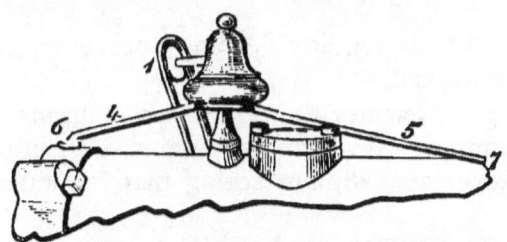

FIG. 97.—Arrangement of Otto Oiler.

5, fig. 97, and drops into holes at 6 and 7, the one oiling the piston every time the trunk comes forward, the other oiling the valve by suitable gutters.

The Clerk oiling cup is shown at fig. 98; it is not automatic. The screw pin 1 is set in a position marked for each cup, the motor cylinder cup giving 15 drops per minute, and the valve cup 5 drops per minute.

In both Otto and Clerk engines the slide valves should be taken out and cleaned once a week. The charred oil should be carefully scraped out of the gas gutters and igniting ports; the piston also should be drawn occasionally, once in three months being sufficient. The interior of the cylinder should then be cleaned, especially the explosion space. The Otto exhaust valve should be taken out every week and cleaned. The Clerk upper

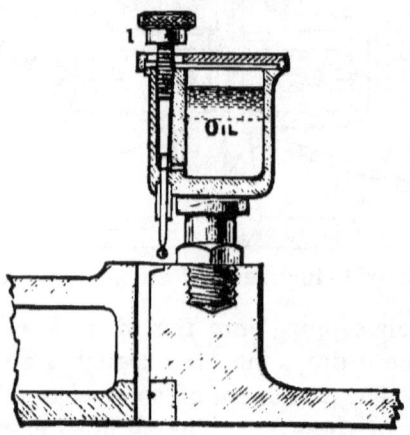

FIG. 98.—Clerk Oil Cup.

and lower lift valves require cleaning once every month if the engine is hard worked.

In working gas engines the two points requiring attention are oiling and cleaning. Never run the engine, without oil, and clean regularly. Never start without seeing that the water circulation is open.

Starting Gear.—Till very lately, gas engines of every power were started by manual labour; in small machines the inconvenience is not great, but with large engines such as those giving from 20 to 50 indicated horses when at full power, the friction is so considerable that difficulties arise. It is difficult to reduce friction so much that a large machine may be turned with sufficient velocity by a couple of men, to get a sure and easy start.

The Brayton petroleum engine was the first to use reservoirs for retaining sufficient air for starting, but they were so faultily constructed, that leakage and loss were so frequent that the apparatus was of little use. Many arrangements have been described by inventors, but no starting gear found its way into public use till that invented by the present author in the end of 1883. The Clerk engine was the first to use starting gear in public, at the

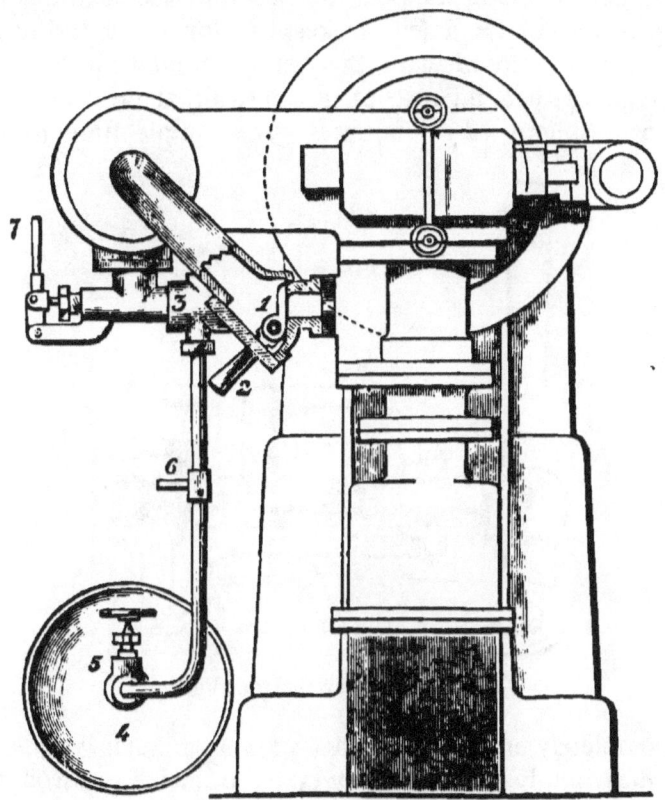

FIG. 99.—Clerk Starting Gear.

beginning of 1884. Since then over 100 engines have been fitted and are at daily work with it.

The Otto engine speedily followed Clerk's in the application of gear, and after them came Tangye and Atkinson.

Starting Gear—Clerk Engine.—The starting gear used in the Clerk engine is shown at figs. 99 and 100. Its action is as follows.

The flap, valve 1, in the communicating pipe between the displacer and motor cylinders is closed, while the engine is running, by the handle 2; the gases in the displacer are thus prevented from entering the motor cylinder, and are compressed through the valve 3, which is an automatic lift, into the reservoir 4, by the stop-valve 5, which must of course be open.

The ignition being stopped, the speed of the engine falls, and the flap is opened for a few strokes, to allow the speed to get up again. It is then closed again, this being repeated till the reservoir 4 is charged with a mixture of gas and air at 60 lbs. per sq. in. above atmosphere. Five minutes gives ample time to charge

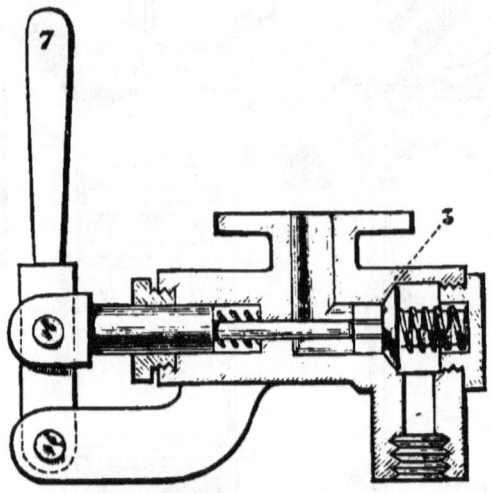

FIG. 100.—Clerk Starting Valve.

from completely empty to 60 lbs. Three minutes suffice if the charge has not been completely taken from the reservoir during the previous start. The relief valve at 6 prevents charging above 60 lbs. per sq. in., the excess blowing into the exhaust pipe. When the reservoir is charged the stop valve 5 is screwed down and the charge is retained in the reservoir till wanted. The reservoir is made of steel, the sides being ½ in. thick and the ends ¾; it is welded throughout, and is tested before leaving the works at 1000 lbs. per sq. in.

The screw down valve 5 and the joint where it is screwed

into the end form the only joints for loss by leakage; numerous joints must be avoided, as it is often necessary to leave the reservoir charged for weeks; the faintest leakage would in so long a time lose the contents, and so the start would require to be made by hand.

The reservoir is so pressure tight, when made as described,

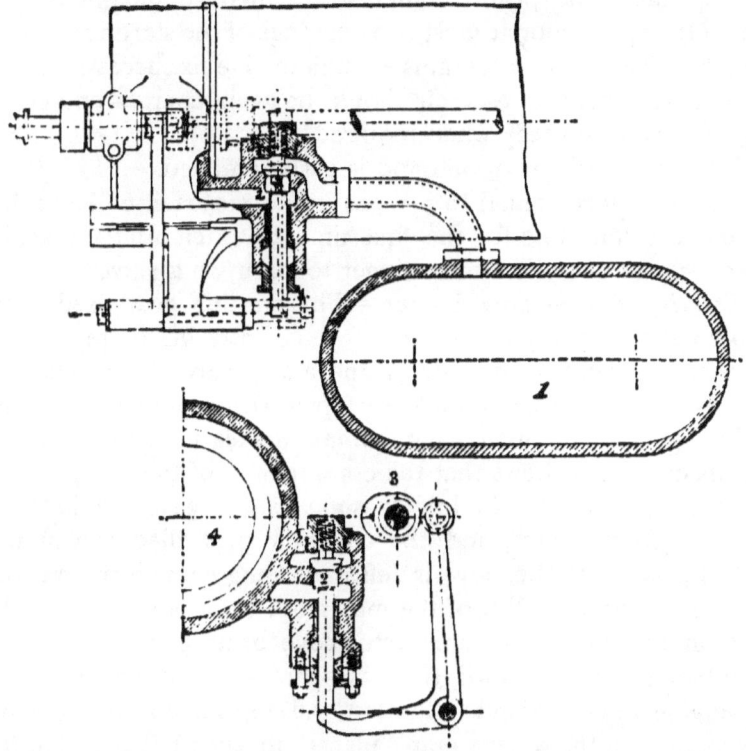

FIG. 101.—Otto Starting Gear.

that the author has left one standing for six weeks and started the engine with ease with what remained.

The starting is effected as follows:

The engine is placed in such position that the motor crank is on the full in centre. The displacer is therefore half forward, the reservoir stop valve is opened, the Bunsen burner is lit, and the gas cock of the engine set at the starting mark. The starting handle 7 is then moved in, opening the valve 3, fig. 100, the gases

R

entering press forward the displacer piston and fill the compression space of the engine, pressing forward the motor piston when its crank comes off the centre. The starting handle is then let go, and the motor piston runs over its ports discharging the contents both of motor and displacer to atmosphere. The engine has thus received a double impulse, one in each cylinder; it is enough to bring round the piston, compress the mixture and get an ignition. One opening or at most two openings of the starting valve are enough. The reservoir contains enough to give six successive starts.

After starting, the reservoir should be again charged and closed so that it may be ready when required.

The gear works very well and is easily handled.

To those accustomed to see gas engines started by hand, it is somewhat astonishing for the first time to watch a large machine move away at once by a mere finger touch upon a valve.

Starting Gear—Otto Engine.—The starting gear used in the Otto engine is shown at fig. 101. It consists of the reservoir 1, the charging and starting valve 2 and a stop valve. The charging valve is loaded so that it does not open with a pressure less than 40 lbs. per square inch, as it communicates with the compression space 4. It follows that the compression of the charge in the cylinder does not lift it, but as soon as the gases explode, the pressure lifts the valve, and the reservoir gets filled slowly with burned gases. If the valve is left open long enough the pressure will rise to within 40 lbs. of the maximum explosion pressure, that is, about 110 lbs. per square inch above atmosphere. The stop valve being screwed down, the gases are retained ready to start the engine when wanted. To start, the stop valve at the reservoir is opened, and the engine crank placed in such position that it is off the centre and on its impulse stroke. The gases then pass through the valve 2 into the cylinder 4, and the valve 2 closes at the end of the stroke actuated by the cam 3. One impulse is thus given and is repeated at the proper time by the action of the cam 3 upon 2, through the intermediate lever. The pressure required is high, because only one forward movement of the piston is available for every two revolutions of the engine. The twin engine therefore starts more easily than the ordinary type of Otto. This gear also works well; it was patented before the Clerk gear, but was later in being introduced into public use.

CHAPTER X.

THEORIES OF THE ACTION OF THE GASES IN THE MODERN GAS ENGINE.

THE general principles developed in this work explaining the causes of the economy of the modern gas engine were first enunciated by the author in a paper read before the Institution of Civil Engineers in April 1882.[1]

He then classified gas engines in three great groups:

Type 1.—Explosion, acting on piston connected to crank. (No compression.)

Type 2.—Compression, with increase of volume after ignition, but at constant pressure.

Type 3.—Compression, with increase in pressure after ignition, but at constant volume.

It was proved that under comparable conditions the relative theoretic efficiencies of the three types were

$$\text{Type } 1 = 0\cdot21$$
$$\text{Type } 2 = 0\cdot36$$
$$\text{Type } 3 = 0\cdot45$$

It was also shown that in the actual engines the real efficiency could not be so high as the theoretic, mainly because of the large proportion of heat lost through the sides of the cylinder, by the exposure of the flame which filled the cylinder to the comparatively cold enclosing walls. A balance sheet was given showing the disposal of 100 heat units by a compression engine. Of the 100 heat units, 17·83 were converted into indicated work, 29·28 were

---

[1] 'The Theory of the Gas Engine,' by Dugald Clerk : *Minutes, Institute Civil Engineers, London.* Paper No. 1855 April 1882.

discharged with the exhaust gases, and 52·89 units passed through the sides of the cylinder into the water jacket.

The economy of the Otto engine over its predecessors, the Lenoir and Hugon engines, was clearly proved to be due to the fact of its using compression previous to explosion.

These conclusions were very generally accepted by scientific and practical men who had studied the subject, and in February 1884 the late Prof. Fleeming Jenkin, then Professor of Engineering at the University of Edinburgh, delivered a lecture at the Institution of Civil Engineers in London, on 'Gas and Caloric Engines.'[1] He had recalculated the efficiencies due to compression, with the result of corroborating the present writer's conclusions. He states :

'If I were to compress gas to 40 lbs., a pressure which is used not unfrequently, the theoretical efficiency would be 45 per cent. We actually get something like 24 or 23 per cent.; we know that one-half of the heat is taken away by external cooling. Thus we find a very close coincidence between the calculated efficiency of those engines and that which we actually obtain, only we throw away about one-half of the heat in keeping the cylinder cool enough to permit lubrication. If we compress to 80 lbs. we have a theoretical efficiency of 53 per cent. If we do not compress at all, as Mr. Clerk has told you, we have a theoretical efficiency of only 21 per cent., so that we have it in our power to increase the theoretical efficiency very greatly by increasing the pressure of the gas and air before ignition. I have no doubt that the great gain of efficiency in the Clerk and Otto engines is really due to the fact of the compression ; this being done in a workmanlike way and carried to a very considerable point.'

The advantages of compression could not be stated with more clearness and truth.

In the same year there was published in Paris an able work entitled 'Etudes sur les Moteurs à Gaz Tonnant,' by Professor Dr. Aimé Witz, of Lille, in which the theoretic efficiencies of the different types of cycle are calculated for a maximum temperature of explosion of 1600° C., and temperature before explosion of 15° C.

---

[1] 'Heat in its Mechanical Applications': *Institute Civil Engineers' Lectures*. Session 1883-84.

He adopts the same classification as the present writer did in 1882, and finds the efficiencies:

$$\text{Type } 1 = 0\cdot 28$$
$$\text{Type } 2 = 0\cdot 38$$
$$\text{Type } 3 = 0\cdot 44$$

which are almost identical with the author's figures.

He also arrives at the conclusion that compression is the great source of economy in the modern gas engine. At p. 53 he says: 'I find myself again in agreement with Mr. Dugald Clerk when he affirms that the success of Otto is due to compression alone, and not to the extreme dilution of the explosive mixture in the products of the combustion of a precedent explosion.'

He then proceeds to quote from the present writer's paper, and adheres to the statement that—

'Without compression previous to ignition an engine cannot be produced giving power economically and with small bulk.'

Compression previous to ignition gives two great advantages:

(1) A thermodynamic advantage (improved theory of the cycle);

(2) Higher available pressures and smaller cooling surfaces —the joint result being an economy in practice nearly fourfold that of the old non-compression engines.

Mr. Otto's Theory.

Previous to 1882 the nature of the improvement obtained by compression was imperfectly understood, and this notwithstanding the very clear, though qualitative, statements of Schmidt, Million, and Beau de Rochas. An erroneous theory of the cause of the economy of the Otto engine was widely circulated and gained considerable support.

It was enunciated in Mr. Otto's specification of 1876, No. 2081, and it was and is still, so far as the author is aware, supported by men so distinguished as Sir Frederick Bramwell, Dr. Slaby of Berlin, Prof. Dewar of the Royal Institution, and Mr. John Imray.

According to Mr. Otto, all gas engines, previous to his patent of 1876, obtained their power from the explosion of a homogeneous charge of gas and air. By the explosion excessive heat was evolved, and the pressures produced rapidly fell away; the excessive heat was rapidly absorbed by the enclosing cold walls.

This caused great loss and gave very wasteful engines. Two methods were open to obtain better economy:

1st, by using a very rapid expansion, so that the heat had but little time to be dissipated;

2nd, by using slow combustion; that is, by causing the inflammable mixture to evolve its heat slowly, so that the production of excessive temperatures and pressures was avoided.

By the first method all the heat was supposed to be evolved at once, and a high temperature was produced: by the second method the heat was evolved gradually so as to give a low temperature and pressure which was sustained throughout the stroke, and which was advantageously utilised by the piston while moving at a moderate speed. Mr. Otto states that this gradual evolution of heat may be produced by stratifying the charge of gas and air. Instead of using the homogeneous charge of Lenoir and Hugon, Mr. Otto uses a charge which he states is not homogeneous but heterogeneous. He affirms that his invention lies in the method or process of forming this stratified charge in a gas-engine cylinder, and that, in addition to the explosive mixture, there must be present in the cylinder a mass of inert gas which does not burn but which serves to absorb the heat of the explosion and prevent the loss which would otherwise occur by the cooling effect of the cylinder walls.

The 'inert' gas may be either air alone which is capable of supporting combustion, or the products of combustion which are incapable of supporting combustion, or a mixture of both. It is not sufficient that a mere film of this inert gas be present; there must be what is termed a 'notable' quantity.

Mr. Otto proposes to form this heterogeneous or stratified charge by first drawing into the cylinder a charge of air alone; and second, a charge of explosive mixture, or by leaving in the cylinder a sufficient quantity of the products of a previous combustion to form a 'notable' quantity of inert diluent.

The compression space in the Otto engine is supposed to contain a sufficient volume of burned gases to form the inert diluent, so that the whole stroke of the piston is available for taking in the explosive charge.

Suppose the piston to begin its charging stroke: the coal-gas and air mixture flows into the cylinder through the inlet port and mixes to some extent with the inert gas already in the space; but the mixing is incomplete, and at the piston itself the charge is supposed to consist entirely of exhaust gases. So that, while the charge at the igniting port is readily explosive, that at the piston is not explosive at all, and between the igniting port and the piston the composition of the charge varies from point to point.

This 'arrangement of the gases' is supposed to be retained during compression, and exist at the moment of explosion. The compression space contains a 'packed charge,' which consists of an explosive mixture at the one end, and between the explosive mixture and the piston a cushion of inert fluid, which is uninflammable and serves the double purpose of relieving the piston from the shock of explosion and absorbing heat which would otherwise be lost by conduction.

By this device, heat is gradually evolved. The flame originated in the port burns at first with great energy and spreads from one combustible particle to another, more and more slowly as it approaches the piston, where the particles are dispersed more and more in the inert gas. The mixture is so arranged that this burning lasts throughout the whole stroke, and is complete very shortly before the exhaust valve opens.

The entire cylinder is never completely filled with flame, but the charge at one end has burned out before the flame arrives at the other end.

Dr. Slaby comes forward in support of this hypothesis in an interesting report published as an Appendix to Prof. Fleeming Jenkins' lecture already referred to.

Dr. Slaby states: 'The essence of Otto's invention consists in a definite arrangement of the explosive gaseous mixture, in conjunction with inert gas, so as to suppress explosion (and nevertheless insure ignition).

'At the touch hole, where the igniting flame is applied, lies a strong combustible mixture which ignites with certainty. The flame of this strong charge enters the cylinder like a shot, and during the advance of the piston it effects the combustion of the farther layers of dispersed gaseous mixture, whilst the shock is deadened by the cushion of inert gases interposed between the combustible charge and the piston.

'The complete action takes place in a cycle of four piston strokes. The first serves for drawing in the gases in their proper arrangement and mixture; the second compresses the charge; during the third the gases are ignited and expand; and finally, by the fourth the products of combustion are expelled. The essential part of the working is performed by the first of these strokes, by which the charge is drawn in and arranged, first air, then dilute combustible mixture, and finally strong combustible mixture. This arrangement is obtained by the working of the admission slide. Moreover, after discharge of the products of combustion, a portion remains in the clearance space of the cylinder, and this constitutes the inert layer next the piston. By this peculiar arrangement of the gases, the ignition and combustion above described are rendered possible, whilst the products of previous combustion form a cushion, saving the piston from the shock of the explosion of the strongly combustible mixture at the farther end of the cylinder.'

Having stated the essence of Otto's invention, Dr. Slaby proceeds to compare the Otto and Lenoir indicator diagrams, to show that the Otto diagrams prove that the above actions occur in the engine. He finds that the Otto expansion line is somewhat above the adiabatic line, and that the Lenoir expansion line is below it. That is, the Otto diagram gives evidence of heat being added or combustion proceeding in the cylinder during the whole expansion stroke, and the Lenoir diagram gives evidence of loss of heat, not gain, during a similar period. If a mass of expanding gas traces on the diagram the adiabatic line, then it appears as if no loss of heat occurred; but as the temperature of the flame filling the cylinder is known to exceed 1200° C., it must be losing heat to the water jacket. To make the expansion line keep up to the adiabatic a great flow of heat into the gas must be taking place, and as the only source of heat is combustion, it follows that the gas is burning during the expansion period.

Dr. Slaby calculates the proportion of heat evolved by the explosion in the Otto engine as 55 per cent., leaving 45 per cent. to be evolved during expansion.

This he states is due to the portion of the charge which continues to burn after the explosion.

The curve differs from Lenoir's in this, that while in Lenoir's engine *all the heat is evolved at the moment of explosion,* leaving none to be evolved during expansion, in Otto's only a part is evolved at first, and the reserved portion keeps up the temperature during expansion.

He concludes from his experiments that the action of the Otto engine is truly as Mr. Otto states in his specification—explosion is suppressed and a slow evolution of heat is obtained, and this slow evolution of heat is the result of the invention and the cause of the economy of the engine.

In addition to this indirect proof, experiments have been made at Deutz and elsewhere to show directly that stratification has a real existence in the Otto engine.

An Otto engine was constructed, specially fitted with two igniting valves; one valve was placed on the side of the cylinder at the end of the explosion space next the piston, so that it could ignite the gases at the piston; the other valve was the usual one at the end of the cylinder, igniting the gases in the admission port.

Experiments were made to discover if the side valve would fire the mixture at the piston; it was found that it did so. Consecutive ignitions were obtained there.

Diagrams were taken for comparison, with the end and the side valves in alternate action, care being taken to keep the charge in the same proportions during the trials. It was found that although the side valve ignited as regularly as the end valve, yet the diagrams were different. Instead of the usual rapid ascending explosion line, the explosion took place more slowly, and the maximum pressure was not attained till late in the stroke.

The ignitions were slower from the side valve than from the end valve. If an uninflammable cushion, such as Dr. Slaby so clearly describes, existed at the piston, one would expect that the side valve would fail entirely, but it ignited quite regularly although more slowly than the end valve.

This experiment is considered to prove stratification.

To make stratification visible to the eye, a small glass mode was constructed. It consisted of a glass cylinder of about $1\frac{1}{2}$ ins. internal diameter, containing a tightly packed piston connected to a crank; the stroke was about 6 ins.; when full back, the piston left a considerable space to represent the explosion space. A brass cover was fitted to the end of the tube, and in it was bored a hole of about $\frac{3}{4}$ in. diameter, representing the admission port; in this hole was screwed a pet cock to which a cigarette was affixed.

On lighting the cigarette and then moving the piston forward by the crank, it was seen that the smoke of the cigarette which passed in did not completely fill the cylinder; the smoke slowly oozed in and left a large clear space between it and the piston. The smoke was supposed to represent the charge of gas and air rushing in, and the clear air behind the piston the cushion which was said to exist in the Otto engine. It was supposed that in the glass cylinder was repeated on a small scale the action of the gases occurring on a larger scale in the Otto engine. In a recent paper in a German engineering journal, Dr. Slaby recounts this experiment, and lays great weight upon it. He considers that it undoubtedly proves the truth of the Otto theory.

In discussion Mr. John Imray concisely states the Otto position as follows :

'The change which Mr. Otto had introduced, and which rendered the engine a success was this: that instead of burning in the cylinder an explosive mixture of gas and air, he burned it in company with, and arranged in a certain way in respect of, a large volume of incombustible gas which was heated by it, and which diminished the speed of combustion.'

And Mr. Bousfield states it in similar terms :

'In the Otto gas engine the charge, varied from a charge which was an explosive mixture at the point of ignition to a charge which was merely an inert fluid near the piston. When ignition took place, there was an explosion close to the point of ignition

Theories of Action of Gases in Modern Gas Engine 251

that was gradually communicated throughout the mass of the cylinder. As the ignition got further away from the primary point of ignition, the rate of transmission became slower, and if the engine were not worked too fast the ignition should gradually catch up the piston during its travel, all the combustible gas being thus consumed. When the engine was worked properly the rate of ignition and the speed of the engine ought to be so timed that the whole of the gaseous contents of the cylinder should have been burned out and have done their work some little time before the exhaust took place, so that their full effect could be seen in the working of the engine. This was the theory of the Otto engine.'

From these quotations it will be seen that Mr. Otto's supporters agree that Mr. Otto has invented a means of suppressing explosion, and substituting for explosion a regulated combustion, and that this process is the cause of the economy of the engine. They are agreed that he has succeeded in preventing explosion, and that he does this by arranging or stratifying the charge which is to be used. They consider that engines previous to Mr. Otto's were wasteful because they used a homogeneous and therefore explosive charge, and that Mr. Otto's engine is economical because it uses a heterogeneous or stratified charge, which is consequently non-explosive.

Discussion of Mr. Otto's Theory.

The primary fallacy of Mr. Otto's theory lies in the assumption that previous engines were more explosive than his, and that in previous engines all the heat was evolved at once: as a plain matter of fact this is incorrect. In the Lenoir and Hugon engines, as in all explosive engines, little more than one-half of the total heat is evolved by the explosion, and the portion reserved is evolved during the stroke of the engine.

The following test of a Lenoir engine, made by the author in London, very clearly shows the suppression of heat at first:

Lenoir engine rated at one horse power.
Cylinder $7\frac{1}{4}$ inches diameter; stroke $11\frac{3}{4}$ inches.
Average revolutions during test, 85 per minute.
Gas consumed in one hour, 86 cubic feet.

With full load, indicated horse power, 1·17 (average of 9 diagrams).

Gas consumed per indicated horse power per hour, 73·5 cubic feet.

Maximum temperatures of explosion, 1100° to 1200° C.

Mixture in engine 1 vol. coal gas, 12·5 vols. of air and other gases.

Heat evolved by explosion, 60 per cent. of total heat.

The proportion of the mixture was calculated from the points of cut-off on the diagram, and after making allowance for the volume of burned gases in the clearances of the engine. It will be observed that only 60 per cent. of the gas is burned at first, leaving 40 per cent. to be burned during the stroke, and also that the temperature of the explosion never exceeds 1200° C. Now in the Otto engine, according to Thurston, 60 per cent. of the heat is evolved at explosion, and 40 afterwards, and the usual maximum temperature is about 1600° C. So that, so far as the slowness of the explosion is concerned, there is no difference, and in the intensity of the temperature produced, the Otto exceeds the Lenoir.

It is difficult to understand how Dr. Slaby could fall into so obvious an error as he did, and suppose that more heat was kept back in the case of the Otto explosion. At the time he wrote his report, accounts of Hirn's, Bunsen's, and Mallard's experiments on explosion were in existence, all of them agreeing on the fact of a large suppression of heat at the maximum temperature of the explosion, although differing in the explanation of the fact.

Hirn even stated that in the Lenoir engine the pressures fell far short of what should be, if all the heat were evolved at once. Yet Dr. Slaby, in the presence of all this definite and carefully ascertained knowledge, is astonished when he finds only 55 per cent. of the total heat evolved by the explosion in the Otto engine, and the only explanation which occurs to him is that of stratification.

If stratification exists at all in the engine, then it produces no measurable change in the explosion; it neither retards the evolution of heat, nor does it moderate the temperature.

The explosion and expansion curves are precisely what they would have been with a homogeneous charge.

Theories of Action of Gases in Modern Gas Engine 253

The mere fact that heat is suppressed in the Otto explosion proves nothing, because a precisely equivalent amount of heat is suppressed in all gaseous explosions, and Dr. Slaby's contention, based upon the supposed peculiarity of the Otto, falls to the ground.

Dr. Slaby has been led into error by the fact that the expansion line of the Lenoir diagram falls below the adiabatic, while the expansion line of the Otto diagram remains slightly above it or upon it. He assumes that in the Lenoir no heat is being added during expansion, whereas just as much heat is being added, or just as much combustion is proceeding during the Lenoir stroke, only the cooling of the cylinder walls is greater, and the heat is abstracted so rapidly that the line falls below the adiabatic. This is due to two causes, (1) the greater proportional cooling surface exposed by the Lenoir engine, and (2) a longer time of exposure. The absence of compression and the slow piston speed makes the loss greater.

Although quite as much heat is evolved during the stroke, it is overpowered by the greater cooling, and the line falls under the adiabatic. This fall is evidence of greater cooling, not of less evolution of heat.

In a recent paper,[1] 'Die Verbrennung in der Gasmaschine, Professor Schöttler makes this explanation of the difference between the lines, and states that 'Whether stratification exists or does not exist in the Otto engine it is unnecessary, and is not the cause of the slow falling of the expansion line.' In all crucial points the Otto theory breaks down, as proved by diagrams taken from his engine.

The explosion is not suppressed; the maximum temperatures produced are not lower than those previously used; the mixture used is not more diluted than in the previous engines, and the intensity of the pressures, as well as the rate of their application, is greater.

The mixture in the engine from Slaby's figures is 1 vol. coal gas to 10·5 vols. of other gases, and from Thurston's figures 1 vol coal gas to 9·1 vols. of other gases, while Lenoir often used 1 vol. gas to 12 of air.

[1] *Zeitschrift des Vereines deutscher Ingenieure.* Band xxx., Seite 209.

The engine instead of using a less explosive power than the Lenoir engine uses one more intensely explosive.

The effect of the reduction of cooling surface and increase of piston velocity is to diminish the loss of heat to the cylinder walls, and the slowly descending line is not the cause of the economy, but is the effect and evidence of it.

Stratification.—The inquiry into the existence or non-existence of stratification in the cylinder has no practical bearing on the question of economy, as the explosion curves act precisely as they would with homogeneous mixtures. Scientifically, however, the question is interesting and will be shortly considered.

The evidence which it is considered proves its existence in the Otto engine is in the author's opinion most unsatisfactory. Dr. Slaby distinctly asserts the existence of an inert stratum next the piston, 'interposed between the combustible charge and the piston,' and Mr. Imray speaks of the 'arrangement of the charge in respect of a large volume of incombustible gases,' and Mr. Bousfield of 'a charge which was merely an inert fluid next the piston.' Yet all the evidence in support of these positive assertions is given by one experiment made with an Otto engine, and one with a small glass model. The evidence given by the experiment on the engine itself, in the author's opinion, disproves stratification in the Otto sense altogether. If the inert stratum next to the piston had any real existence, then the side igniting valve in the experiment made by Mr. Otto, should not have ignited the mixture at all. The fact that it did ignite regularly and consecutively, proved most distinctly that the gas next the piston was not inert but was explosive, and being explosive in itself it could not act as a cushion to absorb heat or shock. That experiment alone settles the question, and proves at once the visionary nature of the cushion of inert gas next the piston.

The fact that the ignitions were slower than those from the end slide does not get rid of the fact that ignition did take place, and to those who understand the sensitive nature of any igniting valve, it will not be difficult to comprehend how small a difference in adjustment will cause late and slow ignitions. At the very utmost the experiment points to a small difference in the dilution of the explosive mixture at the piston and that at the end port.

Experiments made by the author also prove that the mixture in the Otto cylinder is present in explosive proportions close up to the piston. The piston of a $3\frac{1}{2}$ HP Otto engine was bored and fitted with a screw plug, which carried a small spiral of platinum wire in electrical connection with a battery; the platinum spiral projected from the inner surface of the piston by a quarter of an inch. When the engine was running in the usual way, the wire was made incandescent by the battery and the external light was put out. It was proved that by a little care in getting the platinum to a certain temperature, the engine worked as usual, igniting regularly and consecutively. The spiral was made just hot enough to ignite when compression was complete, but not hot enough to ignite before compressing. If an incombustible stratum had existed even so close to the piston as $\frac{1}{4}$ in. then the wire should never have been able to ignite the charge at all. If the wire was made too hot, then ignition often took place while the charge was still entering, proving that no stratification existed even while the charge was incomplete. A little consideration of the arrangement of the Otto engine will show that stratification cannot have any existence in it. The end of the combustion space is usually flat, and sometimes the admission port projects slightly into it; the area of the admission port is about $\frac{1}{30}$ of the piston area; accordingly the entering gases flow into the cylinder at a velocity thirty times the piston velocity, or at the Otto piston speed, about 120 miles an hour.

Great commotion inevitably occurs; the entering jet projects itself through the gases right up against the piston, and then returns eddying and whirling till it mixes thoroughly with whatever may be in the cylinder. The mixture becomes practically homogeneous even before compression commences.

Experiments made by Dr. John Hopkinson and the author on full size glass models of the Otto cylinder show this mixing action very beautifully. A $3\frac{1}{2}$ HP Otto cylinder was copied in every proportion in glass, and the valve was so arranged that it passed a charge of smoke at the proper time. The piston was placed at the end of its stroke, leaving the compression space filled with air. When pulled forward the valve opened to a chamber filled with smoke, and the smoke rushed through the port, projected right through the air in the space, struck the piston, and filled the cylinder uniformly, much faster than the eye could follow it. It mixed instantaneously with the air in the cylinder without evincing the slightest tendency to arrange itself in the manner imagined by Mr. Otto. Mr. Otto's experiment with a cigarette and glass cylinder does not, in the most remote degree, imitate the conditions occurring in his engine; the proportions are quite wrong. The model is much too small, and the glass cylinder is too long in proportion to its diameter; then the gases are so badly throttled by passing through the cigarette, that when the piston is moved forward it leaves a partial vacuum behind it, and only a little smoke enters, not nearly enough to follow up the piston, but only sufficient to ooze into the back of the cylinder while the piston moves forward and expands the air which is already in the cylinder. It was easy for Mr. Otto to have copied his cylinder and valve full size and imitated precisely the conditions existing in his engines.

Had he done this he would have proved complete mixing instead of stratification. Why did he refrain from doing this? The question at issue is not, Can stratification be obtained by a specially devised form of apparatus—no one doubts that it can—but, Does stratification exist in the Otto engine? If it does not exist in the Otto engine then it is perfectly plain that it cannot be the cause of the economy of the motor, and it is quite certain that it cannot exist in the Otto engine. Prof. Schöttler, in the paper already referred to, also arrives at the conclusion that stratification has no existence in the Otto engine, and that Mr. Otto's small glass model does not truly represent the actions occurring in the engine.

In all gas engines, when the charge enters the cylinder through a port the residual gases in the port are swept into the cylinder, and while the port itself is filled with gas and air mixture, free from admixture with residual gases, the cylinder contains the gas and air mixture diluted with whatever residual gases exist in the engine which have not been expelled by the piston. The mixture in the port is accordingly stronger and more inflammable than the mixture in the cylinder.

In the Lenoir and Hugon engines this occurred to a marked

extent; in the Hugon engine as much as 30 per cent. of the whole charge consisted of residual gases, and the charge in the cylinder was considerably more dilute than that in the admission port. In the Otto engine this also occurs, but it is not stratification, and it is not a new invention; the cylinder is filled with explosive mixture more dilute than that in the ignition port, but still explosive throughout.

Causes of the Suppression of Heat at Maximum Temperature in Gaseous Explosions.

Although experimenters are unanimously agreed upon the fact of the suppression of heat at the maximum temperatures produced by gaseous explosions, they differ widely in their explanation of the causes producing this suppression.

Three principal theories have been proposed—

1. *Theory of Limit by Cooling.*—This is Hirn's theory, and it assumes that when explosion occurs, a point is reached when the cooling effect of the enclosing walls is so great that heat is abstracted more rapidly than it is evolved by the explosion, and accordingly the temperature ceases to increase and begins to fall.

The maximum temperature falls short of what it would do if no heat were lost during the progress of the explosion to the walls. If it be true that the cold surface of the vessel is the limiting cause, then the maximum pressure produced in exploding the same gaseous mixture, in vessels of different capacity, will greatly vary. When the vessel is small and the surface therefore relatively large, more heat should be abstracted and lower pressure should be produced. This is not the case. The maximum temperature produced by an explosion is almost independent of the capacity of the vessel. Surface does not control maximum temperature, although increased surface increases the rapidity of the fall of temperature after the point of maximum temperature.

2. *Theory of Limit by Dissociation.*—This is Bunsen's theory, and it is undoubtedly largely true. The fact that no unlimited temperature can be attained by combustion, even when the use of non-conducting materials prevents cooling almost completely, is so conclusively established by science and practice that gradual

combustion due to dissociation may be safely taken as occurring to a considerable extent at the higher temperatures used in gas engines. But there is a difficulty in its application to all cases. In experiments made with explosions in a closed vessel the suppression of heat is almost the same at low temperatures as at high temperatures; thus with hydrogen mixtures—

Max. temp. of explosion   900° C. ; apparent evolution of heat 55 per cent.
,,   ,,   ,,   1700° C.   ,,   ,,   ,, 54   ,,

If dissociation were the sole cause, then as water must dissociate more at the higher temperature than at the lower, the apparent evolution of heat should be less at 1700° C. than at 900° C. It is not so. Some other cause than dissociation must therefore be acting to check the increase of temperature so powerfully at 900° C.

3. *Theory of Limit by the Increasing Specific Heat of the Heated Gases.*—Messrs. Mallard and Le Chatelier have advanced the theory that up to temperatures of about 1800° C. dissociation does not act at all or only to a trifling extent. They consider that the gases are completely combined or burned at the maximum temperature of the explosion. But the specific heat of nitrogen, oxygen, and the products of combustion increases with increasing temperature, becoming nearly doubled when approaching 2000° C. The apparent limit is due, not to the suppression of combustion as required by the dissociation theory, nor to the loss of heat by the theory of cooling, but to the absorption of the heat which is completely evolved by the increasing capacity for heat of the ignited gases. The same objection applies to this as to the dissociation theory. If it were entirely true that specific heat increased with increasing temperature, a greater proportion of heat would apparently be evolved at the lower temperatures, which is not always the case.

It is impossible to discriminate between the effect produced by increased specific heat and the effect produced by dissociation on the explosion curves.

Those are the three principal theories which have been proposed, and in the author's opinion none of them completely explains the facts. The phenomena of explosion are very complex, and no single cause explains the limit and other phenomena of gaseous explosion. These phenomena are more complex than have generally been supposed. In many chemical combinations it has been proved by Messrs. Vernon-Harcourt and Esson, and Dr. E. J. Mills and Dr. Gladstone, that the rate at which the reaction proceeds depends upon the proportions existing between the masses of the acting substances present, and those neutral to the reaction, and that combination proceeds more slowly as dilution increases. From this it follows, that in a combination where no diluent is present, the first part of the action is more rapid than the last; at first all the molecules in contact are active, but after some combination has occurred the product acts as a diluent. The last portion of the reaction, having to proceed in the presence of the greatest dilution, is comparatively slow. Such an action the author considers occurs in all gaseous explosions, and is one of the causes preventing the complete evolution of all the heat present at the moment of the explosion.

The subject is a difficult one, and more experiment is required for its complete settlement.

CHAPTER XI.

THE FUTURE OF THE GAS ENGINE.

SINCE 1860, when by the genius and perseverance of M. Lenoir the gas engine first emerged from the purely experimental stage, it has steadily and continually increased in public favour and usefulness. At first more wasteful of heat than the steam engine, it is now more economical; at first delicate and troublesome in the extreme, it is now firmly established as a convenient, safe and reliable motor; at first only available for small and trifling powers, now really large and powerful motors are used in thousands. Many inventors have contributed to its progress, but its present position is in the main due to the patience, energy and commanding ability of one man—Mr. Otto.

In 1860, the efficiency of the gas engine was only 4 per cent.; in 1886, the efficiency of the best compression engines is 18 per cent.

That is, at first a gas engine could only convert 4 out of every 100 heat units given to it into mechanical work, as developed in the motor cylinder; now it can give 18 out of every 100 units as indicated work.

Having advanced in economy more than fourfold in the past twenty-five years, what limits exist to check its progress in the future?

Apart from the greater perfection of the mechanical arrangements of the gas engines of to-day, the great cause of improvement since 1860 is the successful introduction of the compression principle.

Can this principle be much further extended in its application? In the author's opinion, No.

By undue increase of compression the negative work of the engine would be much increased, and the strains would become so great that heavier and more bulky engines would be required for any given power. Friction, due to this, increases more rapidly than efficiency; consequently the gain in indicated efficiency would be more than compensated by loss of effective power. Improvement must be sought elsewhere.

The most obviously weak point of the present engine is insufficient expansion. In the Otto engine the exhaust valve opens while the gases in the cylinder are still at a pressure of 30 pounds per square inch above atmosphere; in the Clerk engine the pressure is sometimes as high as 35 pounds per square inch above atmosphere at the moment of exhaust.

The gas engines discharge pressures, without utilising them, with which many steam engines commence.

There is evident waste here, which can be remedied by using further expansion. In continuing expansion the loss of heat to the cylinder would not be so great as in the earlier part of the diagram, because the temperature is greatly reduced; it may therefore be supposed, without appreciable error, that the added portion of the diagram would give at least as good a result when compared with its theoretical efficiency as the earlier part. If the expansion be carried so far that the pressure falls to atmosphere, then the theoretical efficiency of an Otto engine would be 0·5; theoretically its cycle would then be able to convert 50 per cent. of the heat given to it into indicated work; practically the compression gas engine at present converts one-half of what theory allows; therefore with the greater expansion it may be expected to give one-half of 50 per cent.—that is, expansion only will raise the practical efficiency from 18 per cent. to 25 per cent.

By complete expansion to atmosphere, the gas consumption of an Otto or Clerk engine could be reduced from 20 cubic feet per IHP hour, to 14·5 cubic feet per IHP hour. There are, of course, practical difficulties in the way of expanding, but they will be overcome in time. Mr. Otto has attempted greater expansion in various ways, and so has the author, but as yet neither has succeeded in carrying it beyond the experimental stage.

It must not be supposed, as it too often is, that a high exhaust pressure means an uneconomical engine, or that comparisons of pressure of exhaust give the smallest clues to the relative economy of engines. It is a very common, but a very erroneous, belief that if the pressure in the cylinder of a gas engine is very near atmospheric pressure when the exhaust valves open, that fact is a proof that the engine is economical.

This is not so—indeed, it may be the very reverse.

In engines of type 3, for example, in which, as in the Otto and Clerk engines, the expansion after explosion is carried to the initial volume existing before explosion and no further, it has already been shown that the actual indicated efficiency is quite independent of the increase of temperature above the temperature of compression. That is, the temperature of the explosion may be anything whatever above the temperature of compression without either increasing or diminishing the indicated economy.

Suppose an Otto diagram with three expansion lines, (1) max. temp. 600° C., (2) max. temp. 1000° C., and (3) max. temp. 1600° C., the maximum temperatures in the three cases being attained at the beginning of the stroke, the efficiency of these three lines is identical. Of course the total indicated power increases with increase of temperature, and diminishes with diminution of temperature, but the proportions of the heat given by the engine as work in the three cases remain constant.

The same thing applies to any number of intermediate temperatures.

It might be supposed that the line 1 by expanding more nearly to atmosphere would be the more economical, and that the line 3, because of the high pressure of exhaust, was the more wasteful.

It is a peculiarity of this cycle, with the expansion stated, that the efficiency is absolutely dependent upon compression alone—that is, the ratio of volume before and after expansion—and is quite independent of the maximum temperature.

The case at once alters if expansion be carried to atmosphere. Here the line 3 would give far greater economy than the others, and efficiency would increase with increase of explosion temperature.

Suppose complete expansion successfully applied to the gas engine, and an actual indicated efficiency of 25 per cent. attained, can any further improvement be hoped for?

What causes the difference still existing between theory, which shows a possible 50 per cent., and practice, which may now realise 25 per cent.?

The great loss is heat flowing from the exploded gases through the cylinder walls. Dr. Slaby's balance-sheet of the Otto engine shows—

	Per cent.
Work indicated in cylinder	16·0
Heat lost to cylinder walls	51·0
Heat carried away by exhaust	31·0
Heat lost by radiation, etc.	2·0
	100

By expanding as described it would be altered as follows:

	Per cent.
Work indicated in cylinder	25·0
Heat lost to cylinder walls	51·0
Heat carried away by exhaust	22·0
Radiated loss, etc.	2·0
	100

The work done will be increased by diminishing the loss of heat with the exhaust gases, but the loss of heat to the cylinder walls will remain constant. This assumes, of course, that the increased time of expansion is balanced in loss to cylinder walls by more rapid rate of fall; if the piston velocity is not increased the result will not be quite so good. If, for instance, the piston velocity is constant, and the volume to surface ratio is constant, the expansion will only give results as follows:

Work indicated in cylinder	21·0
Heat lost to cylinder walls and radiated . .	66·5
Heat carried away by exhaust	12·5
	100·0

Expansion so arranged as to be equivalent to the same time of present piston stroke, 0·2 seconds, by increasing piston velocity and rearranging cooling surfaces, will give 25 per cent. of total heat in indicated work: if surfaces and piston speed remain unaltered, so that the time of exposure increases in same ratio as expansion, then 21 per cent. only will be attained. With proper expansion, the loss of heat by the exhaust gases discharging at a high temperature may be greatly diminished, and the efficiency would be increased, but the change would not affect the loss of heat to cylinder walls; it would even increase it.

How can this, the greatest loss in the gas engine, be reduced? The loss depends, as has already been stated, upon the ratio of surface to volume of gases exposed to cooling, upon the time of exposure, and upon the elevation of the temperature of the hot gas above the enclosing surfaces cooling it.

It is evident that as engines increase in power, the capacity of cylinders of similar proportions increase as the cube of the diameter, while the area of the enclosing cold surfaces increases as the square of the diameter. As engines of greater and greater power are constructed, the surface exposed in proportion to volume becomes less and less; the loss of heat from this cause will, therefore, diminish

Increase in piston velocity will also diminish loss, by diminishing time of contact: 300 feet per minute is the usual speed at present, and it cannot be advantageously increased in small engines, as the reciprocations of the parts become too frequent for durability: but in large engines with diminishing reciprocation, the piston speed may be increased to 600 feet per minute, and still be within the limits practised in steam engines.

Increase in temperature of cylinder walls is also advantageous within certain limits. The author has found a difference of as much as 10 per cent. upon the consumption of gas of an Otto engine when at 17° C., and so hot that the water in the jacket was just short of boiling 96° C. It is probable that still higher temperature could be advantageously used, but there is a limit imposed both by theory and practice.

However, the cycle could be modified to permit the use of very hot walls, enclosing the gases at 500° C.

When all these precautions against loss are practised in large engines, and the heat loss is greatly reduced, another complication steps in, which modifies the theory of the engine very considerably. That complication is the property possessed by all explosive gaseous mixtures of suppressing part of their heat— the phenomenon of Dissociation, the 'Nachbrennen' of the Germans, or the apparent change of specific heat or continued combustion of the French and the English.

Although a gaseous explosion expanding in a cold cylinder behind a piston doing work very nearly follows the adiabatic line, yet if expanded under such circumstances that the loss of heat was greatly diminished, it would no longer do so.

In large engines the expansion curve is always above the adiabatic; in small engines it is below the adiabatic.

In fig. 53, diagram taken by Professor Thurston, if all loss of heat to the cylinder could have been prevented, the expanding line would have been an isothermal, the maximum temperature of 1657° would have been sustained to the end of the stroke, and the actual efficiency of the diagram would have been 0·40, that is, 40 per cent.

At the point 7 the temperature would be 1657°, and the gases would still contain 60 per cent. of all the heat given to them, and if expanded to atmosphere adiabatically, the combustion being supposed complete, then 13 per cent. would be added, making a total efficiency of 53 per cent.

If the loss of heat through the cylinder could be totally suppressed, the possible efficiency, taking into consideration the properties of explosive gases is 53 per cent. It is impossible to completely avoid loss to the cylinder, but it will doubtless be greatly reduced.

The united effect of expansion, greater piston speed and reduction of loss of heat to the cylinder by using hot liners, when carried out in an engine of considerable power, would cause the attainment of a practical heat efficiency of at least 40 per cent., and this without any great change in the construction of gas engines now made.

Now, how do these efficiencies compare with those of the steam engine? It is generally admitted that the best steam engines of considerable powers and of the latest type, when in ordinary work do not give an efficiency greater than 10 per cent., that is, they do not convert more than 10 per cent. of the heat given to the boiler in the form of fuel, into indicated work. In small engines of such powers as are comparable with the largest gas engines yet constructed, the results are not nearly so good, an efficiency of 4 per cent. being a good result.

The reader will remember that the term efficiency, as used in this work throughout, is defined to mean the proportion of heat converted into work, to total heat given to the heat engine.

Efficiency is often used in another sense, and considerable

confusion has arisen because of its use in different senses by different writers. In comparing engines differing in their nature, the only standard of comparison possible is the total heat or total fuel given to each engine, and the proportion of total heat or total fuel which that engine can convert into work. The source of power is always combustion, and the temperature of combustion may always be supposed to be the superior limit of temperature whatever the working process, whether steam or air is the working fluid. From the fact of taking the total heat as the basis of comparison, the reader is not to infer that it is possible even in theory to convert all of it into work. Professor Osborne Reynolds, in a lecture before the Institution of Civil Engineers, stated that this seemed to be a belief popular among engineers; the author does not think that this is so.

Certainly, the second law of Thermodynamics is not so widely understood among engineers as it should be, but still, few suppose that it is even theoretically possible to convert all the heat given to an engine into work.

In the discussion on the author's paper on 'The Theory of the Gas Engine,' at the Institution of Civil Engineers, considerable confusion arose from the term efficiency being used in different senses by different speakers. Professor Fleeming Jenkin in his lecture very clearly defines the different legitimate uses of the term.

Returning to the comparison of gas and steam engine heat efficiency, the 10 per cent. of the steam engine is probably very nearly as much as can be ever attained; it may be exceeded by using high pressures and great expansion, but it will never be possible to attain anything like 20 per cent. The limits of temperature are such that if the steam cycle were perfect, only 32 per cent. of the whole heat could be converted into work; at the boiler pressures and condenser temperatures used, the theoretical efficiency of the steam engine cycle is within 80 per cent. of the cycle of a perfect engine, that is, the efficiency theoretically possible is $32 \times 0.8 = 25.6$ per cent. In an experiment made by Messrs. B. Donkin & Co. on a 63 HP compound engine, the results as given by Professor Cotterill in his work on the steam engine are as follows:

	Per cent.
Absolute efficiency	11·1
Efficiency of a perfect engine	28·4
Relative efficiency	39·1

The engine received 100 heat units from the boiler as dry steam, and it gave 11·1 units as indicated work in the cylinder. With the pressures and temperatures given, the steam engine cycle, if perfectly carried out, falls short of the cycle of a perfect heat engine between the limits, so that 22·7 per cent. is the maximum efficiency which could be obtained, supposing no other loss than that due to the imperfection of the cycle. The cylinder losses, condensation, incomplete expansion and misapplication of heat, make the actual indicated efficiency 11·1 per cent., so that half has gone. The furnace loss diminishes the absolute efficiency to 9·2 per cent., and it is extremely improbable that improvement can ever increase this to 18 per cent., which is the indicated efficiency of the gas engine as at present.

It is impossible that the steam engine can ever offer an efficiency of 40 per cent., which is quite possible with the gas engine.

What remains to be done, then, in order to make the gas engine compete with steam for really large powers? At present the largest gas engines do not indicate more than 40 HP, and very few are in use so powerful.

The gas engine, although superior in efficiency as a heat engine to the steam engine, is not superior in economy except for small powers, where steam engines are very wasteful and the cost of attendance relatively great.

The unit of heat supplied in the form of coal gas is more costly than the unit of heat supplied in the form of coal. Gas producers are required which will convert the whole of the fuel into gas as readily as steam is produced, and with no greater loss of heat than a boiler has.

Mr. J. E. Dowson's producer is the only one at present in existence giving suitable gas, and it requires the special fuel anthracite.

The use of ordinary fuel has not yet succeeded.

A good gas producer, giving gas usable and free from tar, is much wanted.

But when all this is done, the gas engine remains in some respects inferior to the steam engine. It would then be a great advance in economy, as it is at present much superior as a heat engine, or machine for the conversion of heat into work. But mechanically it would still be inferior to steam.

As a piece of mechanism, the steam engine is almost perfect : it is started, stopped, and regulated in a very perfect manner. Its motion is, in good examples, almost perfectly uniform under variation of load, and but little fly-wheel power is required, because there is little or no negative work.

Its motion is perfectly under control.

The gas engine itself requires much improvement in this respect ; it is a comparatively inferior machine ; at best it receives only one impulse every revolution when at full power, and when under light loads only an occasional impulse.

Means must be found to make it double acting, and to diminish the power of the impulses instead of diminishing their frequency for governing.

Means must also be found to start and stop as in steam engines ; the present starting gear is a step in this direction, but requires development.

All this can and will be done ; it is a matter of time and patience. It can and will be made as mechanically perfect and controllable as the steam engine. Flame and explosion, seemingly so untameable and destructive, have been to a great extent tamed and harnessed in present engines. Experience is growing, by which it will be as easily and certainly directed in the cylinder of an engine as steam is at present. The furnace, at present separated from the engine, will be transferred to the engine itself, and the power required will be generated as required for each stroke, and the system of storing it up in enormous reservoirs—steam boilers—finally abandoned.

The masses of smoke polluting our atmosphere will be entirely abolished so far as motive power is concerned.

The author cannot do better in conclusion than quote the late Professor Fleeming Jenkin, expressing his belief in the future of this form of motor.

'Since that is the case now, and since theory shows that it is possible to increase the efficiency of the actual gas engine two or even threefold, then the conclusion seems irresistible, that gas engines will ultimately supplant the steam engine. The steam engine has been improved nearly as far as possible, but the internal-combustion gas engine can undoubtedly be greatly improved, and must command a brilliant future. I feel it a very great privilege to have been allowed to say this to you, and I say it with the strongest personal conviction.'

APPENDIX.

ADIABATIC AND ISOTHERMAL COMPRESSION OF DRY AIR.

(Professor R. H. Thurston, Journal of Franklin Institute, 1884.)

One hundred volumes of dry air at the atmospheric mean temperature of 15·5° C. and 14·7 lbs. per square inch undergo change of volume without loss or gain of heat. The temperatures and volumes corresponding to various pressures are given. Also the volumes at the various pressures if the temperature remained constant at 15·5° C.

Absolute pressure in lbs. per sq. inch	Temperature of compression in Centigrade degrees	Volume at temperature and pressures preceding	Volume if temperature constant at 15·5°
14·7	15·5	100·0	
15·0	17·26	98·58	98·00
20·0	42·60	80·36	73·50
25·0	64·76	68·59	58·80
30·0	82·10	60·27	49·00
35·0	98·38	54·01	42·00
40·0	113·86	49·13	36·75
45·0	126·54	45·18	32·67
50·0	138·96	41·93	29·40
55·0	150·53	39·19	26·73
60·0	161·38	36·84	24·50
65·0	171·61	34·80	22·62
70·0	181·29	33·02	21·00
75·0	190·49	31·44	19·60
80·0	199·26	30·03	18·38
85·0	207·66	28·77	17·29
90·0	214·71	27·62	16·33
95·0	223·45	26·58	15·47
100·0	230·91	25·63	14·70
125·0	264·66	21·88	11·76
150·0	293·91	19·22	9·80
175·0	319·87	17·23	8·40
200·0	343·31	15·67	7·35
225·0	364·71	14·41	6·53
250·0	411·57	13·38	5·88
300·0	420·34	11·75	4·90
400·0	480·76	9·58	3·90
500·0	531·21	8·17	2·94
600·0	574·93	7·18	2·45
700·0	603·74	6·44	2·10
800·0	648·80	5·86	1·84
900·0	680·85	5·39	1·63
1000	710·49	5·00	1·47
2000	929·67	3·06	0·74

ANALYSIS OF COAL GAS.

(T. Chandler, Watts' 'Dict.' Supp. 3, Part 1.)

	Heidelberg	Bonn	Chemnitz	London Ordinary coal gas	London Cannel gas
	vols.	vols.	vols.	vols.	vols.
Hydrogen, H	44·00	39·80	51·29	46·00	27·70
Marsh gas, CH_4	38·40	43·12	36·45	39·50	50·00
Carbonic oxide, CO	5·73	4·66	4·45	7·50	6·80
Heavy hydrocarbons	7·27	4·75	4·91	3·80	13·00
Nitrogen, N	4·23	4·65	1·41	0·50	0·40
Carbonic acid, CO_2	0·37	3·02	1·08	—	0·10
Water vapour, H_2O	—	—	—	2·00	2·00

ANALYSIS OF LONDON COAL GAS.

(Humpidge.)

	Sample (A)	Sample (B)
	vols.	vols.
Hydrogen, H	50·05	51·24
Marsh gas, CH_4	32·37	35·28
Carbonic oxide, CO	12·89	7·40
Olefines	3·87	3·55
Nitrogen, N	—	2·24
Carbonic acid, CO_2	0·32	0·38

ANALYSIS OF BERLIN AND NEW YORK COAL GAS.

	Berlin	New York Municipal Gas Light Co.
	vols.	vols.
Hydrogen, H	49·75	30·30
Marsh gas, CH_4	32·70	24·30
Carbonic oxide, CO	0·54	26·50
Ethylene, C_2H_4	4·61	15·00
Nitrogen, N	0·68	2·40
Carbonic acid, CO_2	2·50	1·00
Oxygen, O	0·22	0·50

Analysis of Natural Gas from Gas Wells in Pennsylvania.

(*Watts* '*Dict. of Chemistry*,' *Supp.* 3, *Part* 2.)

	Burns Butler Co.'s well	Lechburgh Westmoreland Co.	Harvey Butler Co.
	vols.	vols.	vols.
Carbonic acid, CO_2	0·34	0·35	0·06
Carbonic oxide, CO	trace	0·26	—
Hydrogen, H	6·10	4·79	13·50
Marsh gas, CH_4	75·44	89·65	80·11
Ethylene, C_2H_4	18·12	4·39	5·72
Hydrocarbons composition not stated	—	0·56	—

INDEX.

ABE

ABEL, SIR FREDERIC, on gun cotton explosions, 88
Actual indicated efficiency, 117
Adiabatic line, 40
Air, compression lines for, 40
Air engine, Ericcson's, 24
— — Joule's, 31
— — Rankine on, 24
— — Stirling's, 25
— — Wenham's, 25
Air and gas mixtures :
— — proportion of, 99, 100, 101
— — in Clerk engines, 193, 195
— — in Lenoir engines, 128, 252
— — in Otto engines, 173, 176
— — in Otto and Langen engines, 147
Analysis of coal gas —
  Berlin, 271
  Chemnitz, 271
  Deutz, 172
  Hoboken, 175
  London, 271
  Manchester, 109
  natural gas, 272
  New York, 271
Apparent indicated efficiency, 117
Atmospheric engines—
  Barsanti and Matteucci, 11
  Brown's, 2
  Gillies', 151
  Otto and Langen, 136
  Wenham's, 35
Atkinson's differential engine, 195
Available heat, definition of, 112

BARNETT's compression engines, 5, 6, 9
— igniting cock, 7, 207
Barsanti and Matteucci engine, 11
Beau de Rochas on compression, 17

CLE

Berthelot on calculation of temperatures, 108
— explosion pressures, 106
— — wave, 114
— time of explosion, 114
Berthelot and Vieille, explosion wave, 87, 88
Bischoff engine, 132
Bousfield on stratification, 250
Boyle's law, 38
Brake, tests of :
— — Brayton engine, 157, 159
— — Clerk engine, 191-4
— — Otto engine, 172, 175, 180, 181
— — Otto and Langen engine, 141
Brayton engine, 20, 32, 152
— — tests of, 157, 159
— — ignition, 217
— — governor, 233
— — petroleum pump, 156
Brooks and Steward's trial of Otto engine, 175
Brown's gas vacuum engine, 2
Bunsen corroborates Davy's experiments, 83
— on explosion pressure, 106
— — velocity of flame propagation, 84
— — highest temperature of combustion, 93
— — dissociation, 257

CALORIFIC intensity, 90
— power, 90
Cartridge space in Million's engine, 17
Cayley's proposed engine, 25
Charles' law, 38
Classification of gas engines, 29
Clerk engine, 184
— — tests of, 191-4
— igniting valves, 215, 217, 223
— governor, 233

CLE

Clerk starting gear, 238
Clerk's explosion experiments, 93
Clutch, Otto and Langen, 140
Combustion and explosion, 79
— heat evolved by, 89
— volumes of products, 82
Combining weights, 80
Compression engines—
  Atkinson's, 197
  Barnett's, 5, 6, 9
  Brayton's, 20, 32, 152
  Clerk's, 184
  Million's, 16
  Otto's, 172
  Siemens', 18, 32
  Stockport, 197
  Tangye's (Robson's), 195
Compression, Barnett on, 5
— Beau de Rochas on, 17
— Jenkin on, 244
— Million on, 16
— Schmidt on, 17
— Siemens, proposed by, 17
— Witz on, 244
Critical proportion of gas in mixtures, 83
Cushion of inert gases, 247, 248
Cycles of action, 29–35

DAVY, Sir H., on inflammability, 82
Deville, St. Claire, on dissociation, 92, 93
Deutz coal gas, 172
Diagrams, indicator, Bischoff, 134
— — Brayton, 158, 160
— — Clerk, 192 5
— — Hugon, 132
— — Otto, 177, 179, 181
— — Otto and Langen, 142, 147, 148, 150
— — Lenoir, 124, 125
— — Simon, 164
— perfect theoretical :
— type 1, 43
— — 2, 47
— — 3, 50, 52, 54
— — 1 A, 54
Differential engine, 195
Dilution of mixtures, 83
Dissociation, Deville on, 92, 93
— Bunsen's theory of, 257
— definition of, 92
— Groves on, 92
— Thurston on, 178
Drake's engine, 10
— ignition, 10
Dulong and Petit's law, 90

FRE

EFFICIENCY, definition of, 37
— of perfect heat engine, 39
— of imperfect heat engine, 41
— formulæ, 56, 57
— apparent indicated, 117
— actual indicated, 117
— o gas in explosive mixtures, 112
Efficiencies, table of, 68
— — of Brayton engine, 162
— — of Clerk engine, 192
— — of Lenoir engine, 128
— — of Otto engine, 172, 176
— — of Otto and Langen engine, 141
Electrical ignition, 203, 205
Equivalent, mechanical of heat, 36
Ericcson engine, fuel used, 26
Exhaust gases, temperature of Lenoir, 122
— — — Otto, 173, 176
Expansion of gases by heat, 38
Explosion, 95, 115
— chemical reactions of, 81, 82
— Clerk's apparatus, 95, 96
— combustion and, 79
— observed and calculated pressures, 104, 106
— proportion of heat evolved by, 113
Explosive mixtures, true, 79–82
— — efficiencies of gas in, 112
— — curves of cooling, 97, 98
— — inflammability of, 83
— — pressures produced by, 99–101
— — flame propagation in, 84–89
— — temperature produced by, 107–111
— — volumes of products, 82

FLAME propagation :
— — Berthelot and Vieille on, 87, 88
— — Bunsen on, 84
— — Mallard and Le Chatelier on, 85–87
— temperature of, 93, 94, 108, 109, 110, 111
— theoretical, temperature of, 91
— temperature of, in Lenoir engine, 126
— — Brayton engine, 161
— — Otto and Langen engine, 146–149
— — Otto engine, 177, 179
Free piston engines—
  Barsanti and Matteucci, 11
  Gillies, 151
  Otto and Langen, 10, 136
  Wenham, 35
  type 1 A, 66

Index

FRI

Friction of Brayton engine, 139
— Otto engine, 174
Furnace engine, Cayley, 25
— — Wenham, 25
— loss in cylinder, 112
Future of gas engine, 260

GARRETT'S Governor (Clerk engine), 234
Gas, coal analysis of—
   Berlin, 271
   Chemnitz, 271
   Deutz, 172
   Hoboken, 175
   London, 271
   Manchester, 109
   natural gas, 272
   New York, 271
Gas and air mixtures, explosion of, 99, 100, 101
— — — best proportions of, 101-104
Gas, consumption of by Bischoff engine, 134
— — Brayton engine, 158
— — Clerk engine, 191-194
— — Hugon engine, 132
— — Lenoir engine, 124, 252
— — Otto, 172, 175, 180, 183
— — Otto and Langen, 141
Gas, efficiency of in explosion mixtures, 112, 113
Gay-Lussac's laws, 82
Gillies engine, 151
Governors of Bischoff engine, 226
— Brayton and Lenoir, 233
— Clerk, 234
— Lenoir and Hugon, 226
— Otto, 230, 231, 232
— Otto and Langen, 227
— Tangye (Pinkney), 235

HAUTEFEUILLE'S, Abbé, engine, 1
Heat, available, definition of, 112
— — table of, 113
— balance sheets of Otto engine, 172, 176
— engines, perfect, 39
— — imperfect, 41
— evolved by combustion, 88, 89
— mechanical, equivalent of, 36
— losses in gas engine, 72
— lost through surfaces in Otto, 172, 176
— of compression, 40, 270
— specific of gases, 90
— specific, constant volume, 89, 90
— specific, constant pressure, 89, 90

MIL

Heat, unit, 89
Hirn's experiments on explosion, 104, 105
— theory of limit by cooling, 257
Hoboken, coal gas, 175
Hot air engines, Ericcson's, 25
— — — Joule's, 25
— — — Rankine on, 24
— — — Stirling's, 25
— — — Wenham's, 25
Hugon's engine, 20, 129
— igniting valve, 209
Huyghen's, gunpowder engine, 1
Hydrogen, 80, 82
— mixtures, 84, 86, 87
— heat evolved by, 89

IGNITING arrangements, 202
— — chemical, 225
— — electrical, 203-207
— — flame, 207-221
— — incandescence, 222-224
Imray on stratification, 250, 254
Inert gas, cushion of, 247, 248
Inert diluent, 246, 247
Inflammability, 82
Inflammation, definition of, 99
Indicator diagrams, theoretical, 43, 47, 50, 52, 54
— — actual (see diagrams)
Isothermal line, 40

JACKET, water, use of, 27
Joule, Dr., hot air engine, 31
Jenkin, Prof. Fleeming, on compression, 244
— — on future of gas engine, 268

LANGEN'S, Otto and, engine, 10, 136
Lebon's engine, 5
Lenoir engine, 13, 15, 30, 118
— electrical ignition, 203
Limits of dilution, 83, 100, 101, 226
Limit of heat evolution, 257-9
London, coal gas, 271
Losses in gas engines, 72, 78
Lubrication, 235-8

MALLARD and Le Chatelier's experiments, 85-87
— — theory of limit, 258
Mechanical efficiency, Otto, 174
Million on compression, 17
— gas engine, 17

MIX

Mixtures, true explosive, 79-82
— best, for non-compression engine, 101
— dilute, 83
Mixing valve, Clerk, 187
— — Lenoir, 121
— — Otto, 170

NEUTRAL gases, cushion of, 247, 248
Non-compression engines—
  Barsanti and Matteucci's, 11
  Bischoff's, 132
  Gillies', 151
  Hugon's, 20, 129
  Lenoir's, 118
  Otto and Langen's, 136
  Street's, 1
  Wenham's, 35
  Wright's, 3
Norton, Prof., on Ericcson engine, 26
Notable quantity, 246

OILERS, Otto's, 236
  Clerk's, 238
Otto's engine, 136
— — governor, 228, 232
— — igniting valve, 242
— — starting gear, 242
— — tests of, 172, 175, 180, 181
Otto's theory, 245
Otto and Langen's clutch, 136
— — — engine, 136
— — — engine, M. Tresca on, 145

PACKED charge, 247
Papin's experiments, 1
Petroleum engine, Brayton's, 152
Petroleum pump, 156
Pinkney's governor, 156
Piston velocity—
  in Lenoir engine, 145
  in Otto engine, 172, 175
  in Otto and Langen engine, 144
Pressures and temperature, 38, 107
— produced by explosion, 99-101
— — if no loss existed, 104-105
Products of combustion, 82, 109
— — proportion :
— — in Hugon engine, 131
— — in Lenoir engine, 131
— — in Otto engine, 173

RANKINE on air engine, 24
— — available heat, 112

THE

Rankine on science of thermodynamics, 36
Ratio of air to gas, in explosive mixtures, 99-101
— — — in Clerk's engine, 193, 195
— — — in Lenoir's engine, 128
— — — in Otto's engine, 173, 176
— — — in Otto and Langen's engine, 14
Ratio of compression space :
— — in Clerk's engine, 192
— — in Million's engine, 17
— — in Otto's engine, 172, 175
Robson's engine, 195

SCHMIDT on compression, 16
Schöttler on stratification, 256
— tests of Otto engine, 180
Siemens' proposed compression, 18, 32
Simon's steam gas engine, 32, 163
Slaby on Lenoir engine, 248-9
— on stratification, 247, 248
— tests of Otto engine, 170-174, 180
Slow combustion, Bousfield on, 250
— — Imray on, 250
— — Otto on, 246
— — Slaby on, 247-9
Specific heats of gases, 90
Starting gear, Clerk, 239
— — Otto, 242
Stockport engine, 197
Stratification, Bousfield on, 250
— experiments in support of, 249-250
— fallacy of, 254, 255, 256
— Lenoir on, 16
— Otto on, 246
— Schöttler on, 256
— Slaby on, 247, 248
Street's gas engine pump, 1

TABLE of efficiencies, 68
Tangye engine (Robson's), 195
Temperature of combustion in Brayton engine, 161
— exhaust in Lenoir's engine, 122
— — Otto's engine, 173, 176
Temperatures of explosion, 107-111
— — in Lenoir's engine, 123, 125
— — Otto and Langen engine, 146-149
— — Otto engine, 177, 179
Theoretic efficiencies, 68
Theories of actions in cylinder, 243
Thermodynamics of the gas engine, 36

Thurston's experiments on Otto engine, 175
— on dissociation, 178
Tresca's experiments on Lenoir engine, 123
— — Hugon engine, 132
— — Otto and Langen engine, 141
— theory of Otto and Langen engine, 145
Type, first description of perfect cycle, 29
— second description of perfect cycle, 30
— third description of perfect cycle, 32
— 1 A, description of perfect cycle, 34
— 1 :
  Lenoir engine, 118
  Hugon engine, 129
  Bischoff engine, 132
— 2 :
  Brayton petroleum engine, 152
  Simon engine, 163
— 3 : 165
  Otto engine, 166
  Clerk engine, 184

Type 3 (*continued*)
  Tangye engine, 197
  Stockport engine, 197
  Atkinson engine, 199
— 1 A :
  Otto and Langen engine, 136
  Gillies engine, 151

VACUUM gas engine, 2
Velocity of flame propagation, 85
Volumes and relative weights of gases, 81
— — — of Deutz coal gas, 172
— — — of Hoboken coal gas, 176

WATER jacket, use of, 27
Wave, explosion, 114
Wedding on dissociation, 183
Weights and volumes, relative, of gases, 81
Weights, molecular, of gases, 81, 82
Wenham's engines, 25, 35
Witz on compression, 244
Wright's engine, 3

www.ingramcontent.com/pod-product-compliance
Lightning Source LLC
Chambersburg PA
CBHW031342230426
43670CB00006B/413